STRUCTURE

ESSENCE BOOKS ON BUILDING

General Editor: J. H. Cheetham, ARIBA

Other titles in the Essence Books on Building Series

FISHER: Walls
HALE: Floors
LAUNDER: Foundations
OWEN: Roofs
SMITH: Brickwork

Frontispiece National indoor gymnasium in Tokyo

STRUCTURE

H. Werner Rosenthal Dipl. Ing., FRIBA

Department of Architecture,
The Polytechnic of Central London

MACMILLAN

Published by

THE MACMILLAN PRESS LIMITED

London and Basingstoke
Associated companies in New York
Melbourne Toronto Johannesburg
Dublin and Madras

SBN 333 12994 6

Printed in Great Britain by
The Whitefriars Press Ltd., London and Tonbridge

DEDICATION

To Olive Mary, my wife, who once again had to spend many long
evenings on her own while this book was taking its course.

Preface

This is not just another contribution to the vast number of textbooks on the "Theory of Structures". In these days, no single book can hope to cover the complexities of structural analysis and the almost endless number of structural possibilities. If it did make the attempt, it would be quite inadequate from the point of view of the specialist, while at best leaving the non-specialist with an incomplete picture of the problems encountered in structural engineering.

What matters to the architect and to anyone concerned with the construction of buildings — and what is not often realized by the structural engineer — is not so much the ability to calculate the sizes of structural members, but a clear understanding of their relation to one another, the contribution they make to the overall concept of a building, and the reasons behind their various shapes.

This understanding requires an appreciation of the laws of nature involved and the response of structures to them. The laws of nature have not changed, but what is in a constant process of change is our ability to interpret them and to comprehend the inherent properties of materials. So far as building is concerned, progress in this latter field is very slow.

This book attempts to show the fundamental laws governing structural behaviour. These can be reduced to one single concept; that of *Static Equilibrium*. This underlies all structural form, whether expressed in overall layout or in individual elements. It is a concept which applies quite irrespective of any particular material, the only differences arising being due to varying properties of elasticity and strength.

Because of this approach, the old division between "statics" (especially "graphic statics") and "strength of materials" has been superseded.

In this book, calculations have nowhere been regarded as an end in themselves. Where they do occur, their purpose is merely to illustrate certain thought processes. Formulae are not regarded as magic codes, to be swotted up for examinations and soon forgotten, but as convenient shorthand descriptions of structural behaviour. On no occasion are they derived from purely mathematical reasoning. However, for the sake of completeness and out of fairness to the reader, various appendices give examples of some fundamental computations for the design of basic structural members. These are based on SI notation and include a short conversion table.

In abstracting from the immense field of analysis the essentials of structural action, it is hoped that this book will make both the designer and constructor of buildings stress conscious and more aware of the meaning of structural form. The engineer may not find a great deal of new information, but being pre-occupied with the calculation of structures, he may find it helpful to have presented to him those aspects of theory which mainly control — or should control! — the design process.

In any case, it is hoped that what follows will help to bridge the gap between architect and engineer which unfortunately seems to widen rather than to become narrower. It may also be of help to the interested lay reader who wants to make himself familiar with the structural implications of the buildings of any period.

H. Werner Rosenthal
1972

Foreword

by
J. E. Gordon,

Professor of Materials Technology
University of Reading

Materials, elasticity and structures constitute a curiously under-estimated and badly understood group of subjects. Failures of structures and of materials in artifacts are common and sometimes disastrous and it is obvious that to such people as engineers, architects and industrial designers the matter is of the first importance. Yet it is true that the majority of students find difficulty and indeed, experience an emotional resistance in understanding the problems of strength and deflection.

This resistance does not really arise because of any mathematical obscurity. For one thing, those who are good at mathematics are often just the people who have most trouble with elasticity and for another, as this book shows, much of the subject involves no more than the most elementary algebra and arithmetic. The central difficulty seems to consist of acquiring an imaginative feeling for the loads and stresses in structures. In my opinion, it is just at this level that the subject ought to be taught to the non-specialist.

In the past, there have been few books of this kind. The present volume will therefore be particularly welcome. It will be useful not only to architects but should also fill a gap in the teaching of engineers. It will also do no harm at all to people like furniture designers! All are concerned with structure. Or should be!

The subjective difficulty in apprehending the nature of structures is curious. Animals — birds, cats, squirrels, monkeys and so on — which perch or climb about in trees seldom break them. They seem to have some kind of instinct for the strength of structures which is absent in most human beings, even in so simple a matter as sitting in a chair.

Yet, as the author of this book remarks, our subjective appreciation of many architectural forms is based on considerations of elasticity. In early Doric architecture, the exaggerated entasis of the columns and the swelling of the echinoi suggest Poisson's ratio effects, conveying the idea that the columns are carrying a massive load. Here we have an exaggeration of the elastic effects not dissimilar to that employed by Shepherd when he made the roof-beams for the Hall of Churchill College excessively deep. In both cases there is the feeling either that the load is greater than it really is, or else the structure is more than capable of carrying it.

In the later Gothic forms, we get the reverse effect and we marvel that so slender a structure can safely carry such loads over the centuries although, in fact, the actual stresses are quite moderate.

Just as much Renaissance painting depended for its effect upon a deliberate distortion of perspective, so much architectural effect deliberately plays upon our structural feelings. But of course, one cannot break the rules until one knows what the rules are. My contacts with students in some of the schools of architecture suggest that far too many of them are content to remain ignorant of what keeps their buildings up. This minor problem is usually left to such banausic people as engineers, while architects pursue their lordly sociological objectives!

I do not believe that the duties of the engineers and the architect ought to be wholly separated. Indeed, as a naval architect, I am horrified at the idea of separating them at all. I would suggest to land architects that although we have of late acquired the habit of building ugly ships which not only lose money but quite frequently break in two, a study of H.M.S. *Victory*, of the *Cutty Sark* and of H.M.S. *Belfast* can teach many lessons in the art of combining appearance with function.

The aesthetic problem of buildings in which the structure is trivial (as it may be in small houses) or is concealed (as it is in most large modern buildings) does not seem to be within sight of solution. It is a depressing thought that probably more buildings have been erected and more ground covered with buildings during the last 30 years than in the whole of previous history. And that only a tiny proportion of all this immense effort can possibly cause our hearts to rise by even a single millimetre. Frankly, nearly all of this construction is beastly and nearly all of it is the fault of architects. It is probably high time that the public rose up and lynched them! Having said this, I have to confess that I do not know what the solution is! Clearly however, it is not to be found by ignoring structural considerations.

The problem of relating structural appearance to structural function is almost as old as architecture itself. Certainly it dates from the beginning of the 5th century B.C. But however much one may admire the architect Palladio, for instance, one cannot but agree with Dr Johnson that it is difficult to approve of an engaged column! Much of the difficulty must surely arise because the problem has generally been considered as exclusively an aesthetic one, argued from a position of massive structural ignorance or else from somewhat brash attitudes of crude functionalism. Mr Rosenthal's book makes an important contribution towards closing this gap.

May I end upon an unfashionable note? Some of the difficulty in apprehending the nature of elasticity arises from the sizes of the figures which we handle in the arithmetic. Strains — and indeed deflections — are usually very small and often tiny fractions which are difficult to imagine. There is not much that we can do about this. By contrast, stresses and elastic moduli are in danger of being represented by very large arithmetical figures equally difficult to apprehend. This, however, we can do something about. If we speak of a stress of 10 tons per square inch we can easily imagine 10 tons operating on a square inch of material. Much the same sort of thing applies to kilograms per square centimetre. Ten tons per square inch is 154,800 kilonewtons per square metre. Trying to imagine 154,800,000 newtons operating in a square metre is like trying to believe six impossible things before breakfast! It can be done, but it is unnecessarily difficult!

J. E. Gordon

Acknowledgements

My thanks are due to Peter Ryalls, BSc, DIC, of Ove Arup and Partners for his help and generous advice on many aspects of the text, especially those referring to reinforced concrete design.

I also acknowledge with gratitude the enthusiastic co-operation of John Cocks, who turned my often very rough sketches into clear and presentable drawings.

I must also thank the Japan Information Centre, London for permission to reproduce the Frontispiece, Messrs Beken of Cowes for Figure 1 and the Cement and Concrete Association for Figures 2, 37 and 41.

Contents

Further Aspects of the Theory of Eccentric Pressure
Pre-stressing

Appendix 1 – SI Conversion Table
Appendix 2 – Calculation of a Roof Truss
Appendix 3 – Calculation of Timber Purlin on Truss shown in
                 Appendix 2
Appendix 4 – Calculation of a Timber Box Girder
Appendix 5 – Calculation of a Slender Timber Column
Appendix 6 – "Kern" Dimensions for Hollow Shapes
Appendix 7 – Calculation of a Continuous Reinforced
                 Concrete Floor Slab

1 The meaning of structure

When talking about architecture or building, either consciously or unconsciously, we talk about structure.

Whether it is the half-timber work of a Tudor mansion or the charm of a beamed cottage, the flying buttresses of a cathedral like Notre Dame or Salisbury, or the dramatic trusses of a tithe barn, it is always structure.

Far more often than we realize, structure is the cause of our emotional response to buildings. It may well be that a great deal of the general indifference to modern architecture is caused by the fact that "structure" (that is, the bones and tendons of a building) is no longer recognizable and therefore no longer appeals to us in a direct way. This is due to a greater sophistication in structural engineering; for example, the steel bars which take up the tension in reinforced concrete cannot be seen. A parallel can be found in other forms of engineering where the activating forces are hidden from our eyes. For instance, a sailing ship with its sails reaching out like wings stirs our emotions. We can project ourselves into its action as if it were a living thing, feeling the forces of the wind as we do (Fig. 1). On the other hand, the diesel engine in a power craft, with its hundreds of horsepower concentrated into a small space, remains largely a remote and utilitarian piece of machinery which may excite only the specialist engineer.

New materials, or the more efficient use of traditional materials and a knowledge of their behaviour, have given us a freedom in design which enables us to plan with few restrictions except financial ones. The form of buildings is therefore no longer so closely limited by structural considerations. This has led to a great deal of meaningless form. At the same time, it has also resulted in many new forms undreamed of in previous ages. On a humbler level the appreciation of the laws of nature which govern structural behaviour has enabled us to use our resources with greater economy and a minimum of waste, compatible with the practical requirements of a building.

Up to a point this has always been a natural endeavour. It can be seen, for instance, in the incredibly slender columns and ribs of Gothic architecture. What science has done is to tabulate past experience and observations into data which enable us to forecast

1

Fig. 1. Structure under sail

performance and design accordingly. So, an understanding of the structural behaviour serves two aspects of architecture; a starting point of design and inspiration to form on the one hand, and a means of economical solutions of building problems on the other. Sometimes the two aspects coincide, sometimes they diverge. No general rule can be laid down, but broadly, one can say that the larger the span of a building the more its form is dictated by structural considerations. It is here that economy of means and the best structural solution may coincide (Fig. 2).

Fig. 2. The cantilevered beams of Rome station follow the line of the bending moments

In smaller buildings, considerations of the availability of materials, labour cost and planning requirements are often at variance with the best solution from a purely structural point of view. For instance, for economy in material and in direct response to the laws of statics the joists in a house could be shaped as in Fig. 3. The reasons for this will be understood later, but it does not require much imagination to see the absurdity of such an idea!

Fig. 3.

All structural problems have one basic requirement in common: *the building must stand up!*

So the following chapters will deal with the fundamental principles by which this is achieved and the laws of nature which govern the lay-out of structures and the shape of structural members. We shall therefore confine ourselves to structural requirements only. This book does not deal with *construction*, which is the art of putting the elements of the building together so that in addition to standing up, they also fulfil all the other needs of building such as keeping the weather out, the warmth in, admitting or excluding light and air and generally shelter the great variety of activities for which our buildings are erected.

2 Forces and equilibrium

There is one requirement for the stability of buildings and one only: *equilibrium.*

Buildings and building components are subject to a great many forces which tend to disrupt the equilibrium. The structure must be able to withstand these attacks.

"Forces" as such are intangible. They can only be perceived by the movement they cause. For instance gravity (which is the most important one) causes a downward movement which we perceive as *weight.* In a state of equilibrium an object remains *static;* that is, movement which gravity attempts to create is *arrested.* This concept of a force as movement or change of state is recognized in the SI notation ("Système International") see App. 1, where forces are measured in "newtons" in accordance with Newton's Second Law of Motion. This defines a force as the "acceleration of mass". As the weight is a localized concept, depending on the existing gravitational field, "mass" represents a more objective notation, valid anywhere in space and on our planet. But since building deals with terrestrial conditions involving only minute differences in the gravitational fields, we need not worry too much about such distinctions.

Values may be simply translated. For instance, a force of 1 lb is equivalent to 4.44822 N and since this is about 450 gram, one N is about 100 g (see App. 1). As forces cause movement which we try to arrest, a great deal of structural behaviour can be deduced from observing or visualizing the movement which would occur if certain components which contribute to the general equilibrium of a structure were removed.

In Fig. 4, for example, the missing member would obviously have to pull back, that is, it would be in "tension". In Fig. 5, however, a compression member would be required. Similarly, the piece of

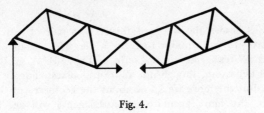

Fig. 4.

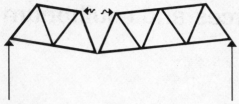

Fig. 5.

wood cut out of the joist in Fig. 6 would have to be in compression, while the missing piece in Fig. 7 would be in tension.

In the girder shown in Fig. 8, the missing diagonal member would have to be in tension, in fact it would meet the shear forces at this point (see p. 73).

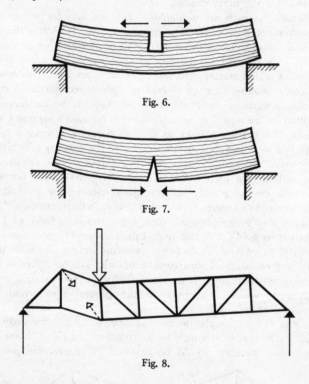

Fig. 6.

Fig. 7.

Fig. 8.

Figure 9 shows the outward thrust of a pair of rafters which would have to be equilibriated by a tie "T", or by a pair of buttresses to resist the turning effect (Fig. 10). A great deal of structural behaviour, though not all, can be assessed qualitatively in this way. But the more we know about the geometry of the action of forces, the more accurate our assessments will be. This also

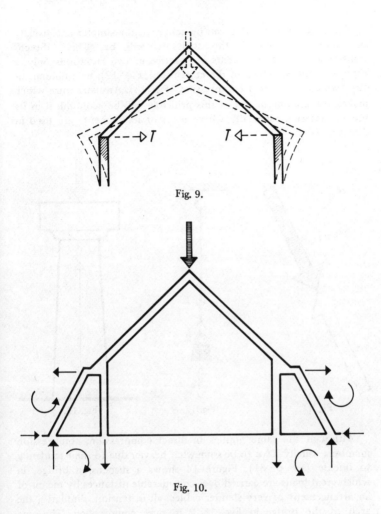

Fig. 9.

Fig. 10.

enables us to draw quantitative conclusions, so that building elements and materials can be arranged so as to fulfil their functions in the best possible way.

Alternatively, such assessments can make us aware of the fact that structural members may not be used to their fullest advantage and this must be weighed against other requirements. So we will deal first with the geometry of the action of forces. There is only one possible state of equilibrium, namely, *the action of a force in one direction which is equilibriated by an equal and opposite force on the same line of action* (Fig. 11). All systems of forces at every point of a structure must ultimately be reduced to this basic state of equilibrium.

7

The more directly this can be achieved, the simpler and usually the more economical the structure will be. This "direct" transmission of forces occurs in practice in two situations only, a rope or cable which is in direct tension (Fig. 12), or a column or strut which is in direct compression (Fig. 13). Any structure which makes the maximum use of this principle will be economical in its use of material, especially where most of its elements are used in direct tension.

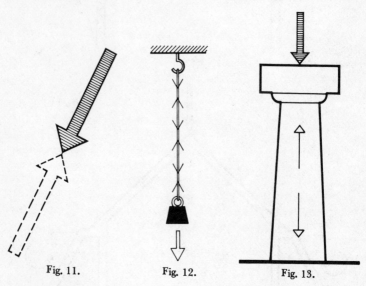

Fig. 11.   Fig. 12.   Fig. 13.

Although the same applies in direct compression, compression members usually have to be somewhat heavier due to their tendency to buckle (see p. 81). Figure 14 shows a suspension bridge, in which great loads are carried over considerable distances by means of an arrangement of very slender cables, all in tension. Similarly, the arch of the bridge in Fig. 15 is in pure compression. The two structures are complementary.

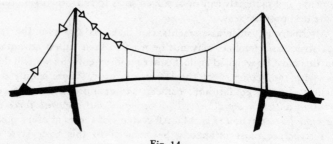

Fig. 14.

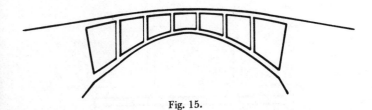

Fig. 15.

The arch is an ancient method of bridging space, while the tension cable has been introduced much more recently, due to the availability of high tensile materials. The mechanics of the two related concepts will be dealt with later.

Unfortunately, in our efforts at bridging space, forces can rarely be equilibrated in this direct way. They are usually forced along routes having considerable "detours" in their travel down to earth and equilibrium (Fig. 16). The dotted line here shows the most direct route but if this were to be adopted, the space (shaded area) would be lost. If the force "W" were split into two equal halves and applied where shown in Fig. 17 the respective "detours" would be smaller (broken lines), the resulting structure lighter and the unusable space in the building less. In other words, the more that loads are distributed, the lighter the supporting structure can be for the same total load.

Concentrated "point loads" should be avoided whenever possible as they lead to considerable structural detours. The worst possible "detour" in this sense is the cantilever (see Fig. 18).

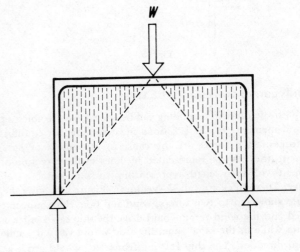

Fig. 16.

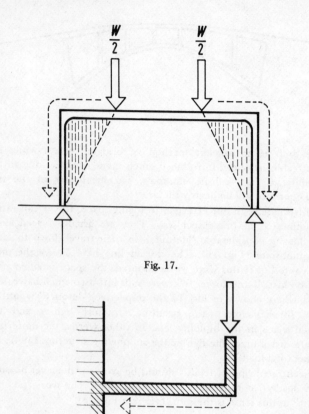

Fig. 17.

Fig. 18.

The polygon of forces and the truss

If the path of forces in a building can be traced, we are able to place structural members in the positions most suitable for equilibrating these forces. So if forces are the cause of movement or *potential* movement, they can be represented by lines in the direction of this movement with their length representing (to some suitable scale) its intensity. These lines are called "vectors". Figure 19 shows the case of a ship subjected to two forces, wind and tide. Here, movement is intended and the wind force would drive the ship say 4 miles in one direction while in the same time, the tide would carry it 1 mile in a different direction. The ship starting from "A" would arrive at "C" and in this case would have covered 4.12 miles. If the ship were a

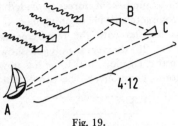

Fig. 19.

"structure" which did not want to move, it would require a force equal and opposite to the one represented by the line A-C in order to "arrest" this movement (Fig. 20). In other words, in the triangle ABC, line AB represents a force of 4 units, say 4 N, the line BC a force of 1N and the line from A to C the resultant movement or *resultant*. The reverse, from C to A, represents the equilibrating force required both in *magnitude* and *direction*.

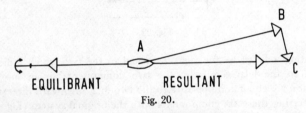

EQUILIBRANT   RESULTANT

Fig. 20.

A system of forces can thus be represented by "vector triangles" and what applies to three forces will equally apply to any number of forces (Fig. 21a, b). In this case, any one force may be the equilibrant to all the others or, in the opposite direction (marked ▷▷), the resultant of any number of forces which are the equilibrants to any of the remainder. Any "resultant" may also be represented by its *components*. In practice this means that any one force may be split into any number of "components", or be

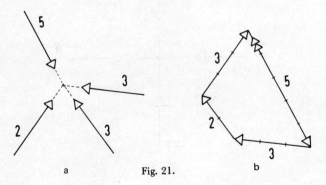

a   Fig. 21.   b

11

equilibrated by any number of forces provided that they can be represented by a complete triangle or polygon.

How does this line up with the requirements for equilibrium stated earlier? In Fig. 22, the vector polygon shown in Fig. 21b has been used again, but each individual force has now been split into a vertical and horizontal component. This shows that two components

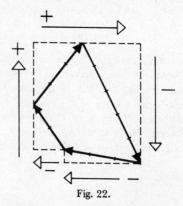

Fig. 22.

indicating a force to the right are equal to the two which indicate forces to the left. Similarly, the two components in an upward direction are equal and opposite to the two in a downward direction. Transferring these six components onto the original system (Fig. 23) shows that the point on which the various forces act would not move in any direction — neither right, left, up nor down. Here, the basic state of equilibrium can be expressed as:

$$\Sigma H = 0 \qquad \Sigma \bar{V} = 0$$

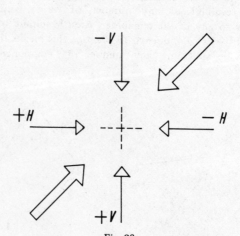

Fig. 23.

12

and if any two forces in this system are represented by their respective resultants, two resultants remain, equal and opposite, and acting on the same line. In short, they cancel each other out.

We have seen that forces are in equilibrium if they are equal and opposite and act along one line. This means that ultimately, they must converge upon one point in space. While this situation is not always immediately apparent, there are numerous building elements in which it becomes obvious. These are usually the simplest and lightest systems for supporting given loads. It will be understood that if any system of forces is in equilibrium, this must apply to each point in such a system.

A case in point was illustrated in Fig. 9. It is shown again in Fig. 24a, b. Here the load "W" is held in equilibrium by the two rafters AB and AC. The forces required in the rafters are represented by the vector triangles ABC where BC represents "W" and AB and AC the equilibrating vectors (Fig. 24b). It can be seen that the rafters tend

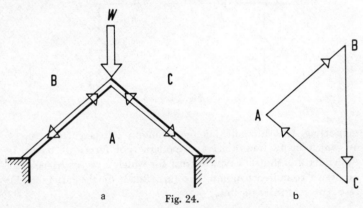

Fig. 24.

to "push up" in their effort to meet the downward force "W". In order to do so they must "push down" at the lower ends and this downward push is met by the vertical force VL and VR and by horizontal forces HL and HR, represented in vector form by the triangle shown in Fig. 25. These would be the buttressing forces shown in Fig. 10.

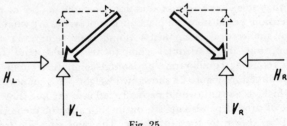

Fig. 25.

13

Instead of such "buttressing", the horizontal forces could also be absorbed by a horizontal tie (Fig. 26). Here, we have the simplest form of a triangulated "truss", in which all loads are transmitted directly along the members either in "compression" (in the rafters) or "tension" (in the tie). Only vertical components are taken by the supports, two walls for example. Otherwise the system is self

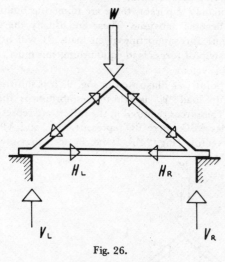

Fig. 26.

supporting. This shows the inherent economy of triangulated trusses, on which many traditional and modern roof forms are based. In Appendix 2 is shown a typical truss for which a vector triangle can be drawn round each point. The three loads on the rafters in this case represent three purlins.

The Funicular line

As mentioned earlier the most direct equilibrium is achieved in direct tension, because in this case very light members or even flexible cables can be used to transmit loads.

This situation is illustrated in Fig. 27. The reader will realize of course that this is the reverse situation from the couple of rafters. Here, tension would take the place of compression. At what is now the top, the vertical components would again require an upward direction, but the horizontal components would tend to pull *inwards* requiring equilibrating horizontal components acting in an *outward* direction. Figure 28 shows two weights "W" supported on a cable which, as a result, would naturally fall into the line shown. The amount of drop depends entirely on the intensity of the horizontal pull "H" as shown by the dotted line. The same situation arises if a

14

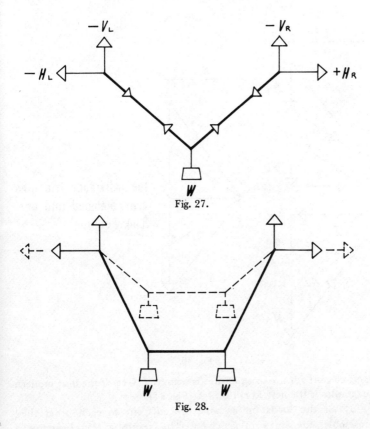

Fig. 27.

Fig. 28.

great number of weights are suspended (Fig. 29). As each point on this line is in equilibrium, vector triangles can be drawn for each point showing exactly the magnitude of the tension in each stretch of cable (Fig. 30a, b). This shows that the first section of cable from point 1 to 2 is under the greatest load. On the whole, a very light

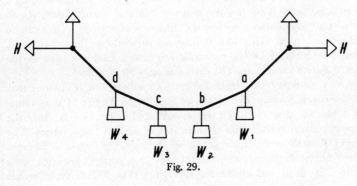

Fig. 29.

15

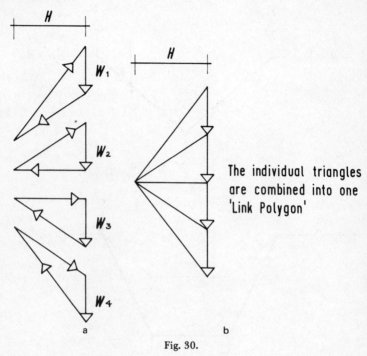

The individual triangles are combined into one 'Link Polygon'

a   b

Fig. 30.

and elegant form of suspension is achieved. One of the best modern examples is the new Forth Bridge (Fig. 31).

If all the loads on a cable are adjacent to each other they resemble the links of a chain. The resulting line becomes a "catenary" line (from Latin *catena*, meaning a chain). This line is very nearly parabolic. As in the general funicular line discussed previously all the stresses remain inside the curve and are either pure tension or pure compression. In Fig. 32 is shown a small chain, like a necklace, and it is obvious that the "drop" depends on the intensity of the tension. Figure 33 shows the line in reverse, representing a parabolic arch. This line becomes a "natural thrust line". As all forces are equilibrated within the thickness of such an arch it can be made very thin. This was first demonstrated by the great Swiss engineer Maillart in his parabolic reinforced concrete shell at the 1939 Zürich exhibition. The shell was only 50 mm thick.

As a catenary can never fall into a semi-circle it is obvious that the classical Roman arch is not an ideal shape. Figure 34 shows how it tends to open up in the region indicated in the sketch. Only the upper part of such a semi-circular arch resembles the catenary. The great Roman arches have stood up through the centuries because their heavy abutments ensure that the "arch action" does not come into play until well above the springings (Fig. 35). Norman arches

16

Fig. 31. Forth road bridge

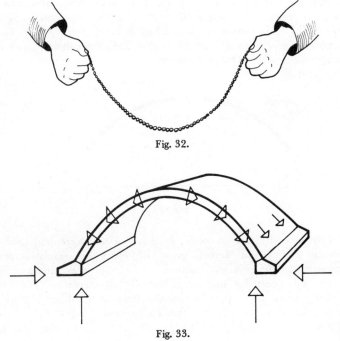

Fig. 32.

Fig. 33.

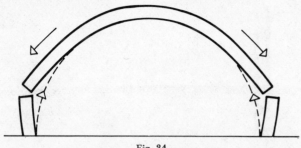

Fig. 34.

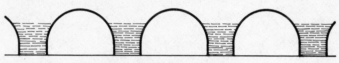

Fig. 35.

also have these heavy abutments. In fact, the top segment of the circle (where the rise equals about one quarter of the span) is near parabolic (Fig. 36). This fact is important in the economics of constructing shells and concrete arches, since it makes possible use of shuttering of constant curvature (that is, circular). Also, when arches or vaults are constructed of identical pre-cast elements as in the case of Nervi's great exhibition hall in Turin (see Fig. 37). This fine roof rises to one quarter of its span and all the members are in compression, assisted by post-tensioning.

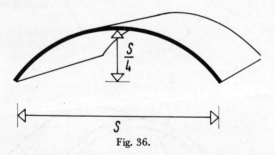

Fig. 36.

Influenced by Roman forms, Renaissance architects used chiefly circular arches and domes, many of which are non-structural according to the principles outlined above. The great dome of St. Paul's in London, for instance, is basically non-structural (Fig. 38). The weight of the lantern is brought down in the direct line by a large hidden cone which is the three dimensional application of the principle shown in Figs 16, 24 and 27. At St. Paul's, the resulting

18

Fig. 37. Exhibition Hall, Turin, Italy (P. L. Nervi)

outward thrust is restrained by a ring chain and the familiar "dome" is supported on a timber framework. On the other hand, the 15th century dome of St. Maria della Fiore in Florence, designed by Brunelleschi, is built up of near-parabolic ribs and is consequently self-supporting (Fig. 39). This Renaissance building is, in fact, almost Gothic in character. Gothic arches and ribs, being nearly parabolic in form, come close to catenaries. Hence their extreme thinness.

The principle of the funicular, with its special version of the catenary shape, has been applied in recent years to a number of roofs of extremely long span. These have been made possible by the introduction of high tensile steel, although ordinary mild steel is also

19

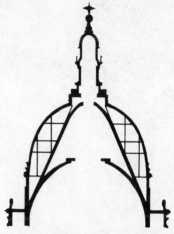

Fig. 38.

used. For instance, a network of reinforcement wire is stretched between a system of pillars which act as buttresses in reverse. The whole is then covered by a thin coating of concrete, usually applied by spraying. An example of this type of structure, used for a large swimming pool in Germany, is shown in Fig. 40.

In Fig. 41 is shown a marketing hall in California, where the tension is balanced by two huge cantilevered hoops which in themselves act as inclined arches. The resulting roof is structurally three dimensional and its stiffness is further increased by the saddle shape. The hoop arches are parabolic and therefore in almost pure compression.

Structures of this kind are extremely light for the long spans involved. The only drawback is a tendency to flutter due to wind suction. This is often provided against by introducing integral window mullions which, although in theory unstressed, in practice take up any tension caused by upward suction.

Reinforced concrete parabolic shapes, especially of the three-dimensional kind, cause shuttering problems. Although they often represent very elegant solutions to the problems of bridging space, they can be very expensive due to high labour costs. The simplest way of obtaining a saddle shaped parabolic roof is by the use of the *hyperbolic paraboloid*. The shape is "generated" by straightline planking or cables extended between two opposing slopes (Fig. 42). The resulting surface shows a typical saddle shape which represents two parabolic curvatures at right angles to each other. One is in compression and demonstrates arch action (A–A). The other is a tensile catenary (B–B). Any horizontal section cut through this shape produces hyperbolic curves, hence the name.

20

Fig. 39. The dome of Florence Cathedral by Brunelleschi is pure structure

Fig. 40. Swimming pool in Germany — a good example of "funicular" roof in reinforced concrete

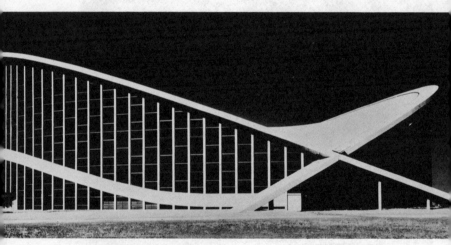

Fig. 41. In the Livestock Hall in South Carolina the concrete hoops essentially support only the upper part of the hyperbolic paraboloid which acts as a catenary in tension

The lower points are usually supported by buttresses. These help to keep the catenary section in shape, restraining it so that it cannot sag. It is also possible to support the top corners only as this would, in itself, keep the arch in shape, but this is rarely done in practice. As mentioned above, there is a certain risk of flutter which can be taken care of by light tension members. The hyperbolic paraboloid can be used in an infinite number of combinations and apart from thin concrete shells it is frequently constructed in plywood.

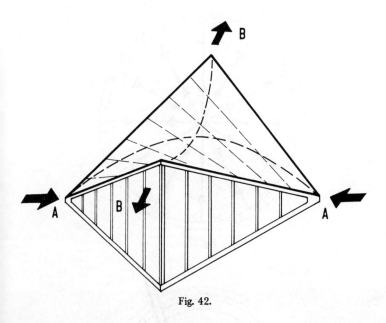

Fig. 42.

Rings and Domes

We have seen that the parabolic or near parabolic curve is the funicular line for parallel loads. In practice, due to wind and other local influences, loads are not always strictly parallel. The traditional bridge arch in masonry, for instance, is thick enough to accommodate the catenary line somewhere within its thickness and in the case of such bridges there is usually adequate buttressing material. In addition, the usually solid masonry above presents any undue deformation.

Catenary lines *can* be circular, however, if the loads are *radial*, either in compression or tension (Fig. 43). This can be seen in the bicycle wheel, in which all the spokes are in radial tension, inducing a circular *compression* (funicular) action in the rim. The tension produces a "pre-stress" condition in the rim, so that any local tension caused by the inevitably occurring uneven loading of the wheel when in use is amply counteracted (Fig. 44). The principle was employed on a large scale in the American pavilion at the Brussels exhibition resulting in an immense circular roof (shown diagrammatically in Fig. 45). This ring or hoop action contributes substantially to the stability of circular domes. The very fact that the dome is in parts non-funicular produces radial thrusts which vary over the extent of each longitude (Fig. 46). In the same way as the hoops of a full barrel keep it in shape (Fig. 47), a dome develops

23

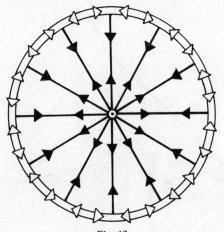

Fig. 43.

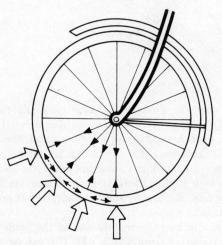

Fig. 44.

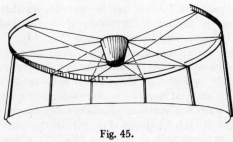

Fig. 45.

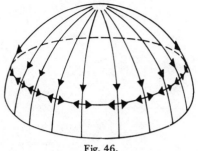

Fig. 46.

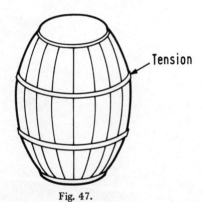

Tension

Fig. 47.

hoop stresses in response to the radial thrust. In other words, each ring of "latitude" becomes a circular "funicular". These are not all in tension, since the lower "latitude" frequently develops compression due to an inward thrust of the longitudes, roughly from a position where they diverge most from the thrust funicular. It is this hoop action which has enabled many masonry domes to survive, as the tensile bond between the masonry elements increased with age.

It is essentially this static system which made Buckminster Fuller's domes possible as the "latitudinal" elements are in tension. The funicular has become one of the most important structural concepts and will be frequently referred to.

3 Moments

Moments in general

We have seen that static equilibrium is achieved in entire structures, or the elements of which they are composed, if forces can be arranged to be equal and opposite, so cancelling each other out. However, in building situations this rarely happens (see Figs 11, 12). In Fig. 48 is shown the typical case of a force (the weight of a person) placed at a considerable distance from a possible equilibrating force (the wall). If a support were placed directly underneath

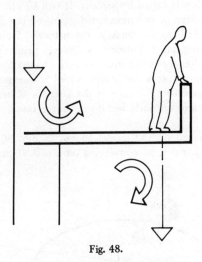

Fig. 48.

the person, as in Fig. 49, equilibrium would be restored and apart from giving this man a somewhat precarious perch, the connecting balcony would be redundant. The balcony represents one method of bridging space.

Wherever space is so bridged, loads and supports fail to coincide, thus leading to the detours referred to earlier. In such cases, forces may be equal and opposite but they *do not act on one line*. The result is a *rotation*, and the curving arrow in Fig. 48 clearly indicates this rotation. No vector triangle can be drawn for this situation and equilibrium can only be restored by a counter-rotation.

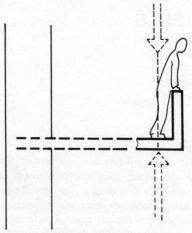

Fig. 49.

Such a situation is called a *moment*. It will be obvious from Fig. 48 that the intensity of this moment depends on the distance of the weight (represented by the man on the balcony) from the support which acts as fulcrum. A moment is directly proportional to such distance (called the *lever arm*) and, of course, to the force itself.

This principle of forces and levers is well known from mechanics but is of the greatest importance in the analysis of structures. So, if forces are equal and opposite but do *not* act in one line (one point), moments occur.

This is illustrated on the model shown in Fig. 50. On a pulley board four weights act, all converging on a nail. If the system is in

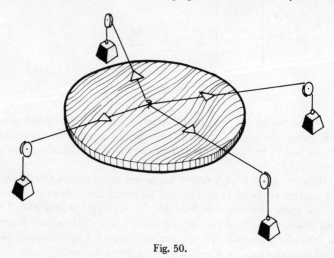

Fig. 50.

equilibrium, each pair can be represented by its resultant as shown in Fig. 23 and the forces will cancel each other out. In the example shown in Fig. 51, the two forces do *not* act on one point and rotation therefore occurs. It would be brought to an end if the pulley board were turned in an anti-clockwise direction until the pulls were opposite.

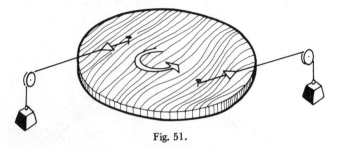

Fig. 51.

In a machine (for instance, a bicycle pedal) this rotation is a desired effect. Such an effect is greatest when the lever arm (that is, the perpendicular distance from the fulcrum), is greatest.

In building, however, we have to arrest rotational as well as any other movement so, further to the statement made earlier, we now have to add

for Equilibrium: $\Sigma M = 0$

or, all moments must add up to zero.

Figure 52 shows this in numerical values. The weight of the wall supplies the equilibrating moments, so that:

+ 80 x 1200 − wt. of wall x 150 = 0.

640 kg

0.150 m

80 kg

1 200 m

720 kg

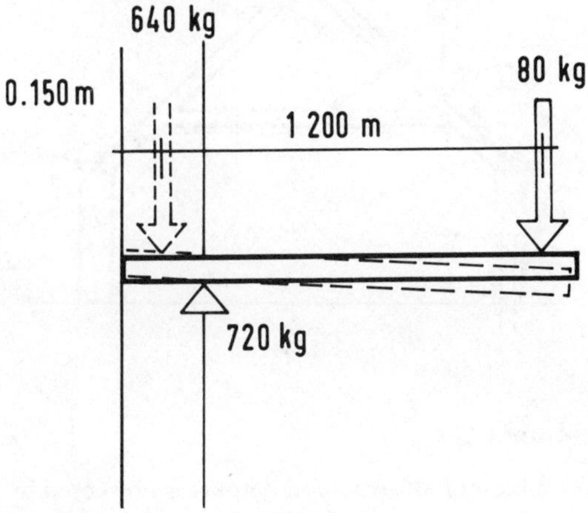

Fig. 52.

The edge of the wall is loaded with 720 kg, so that all upward forces equal all downward forces.

In Fig. 53 is shown the moment set up at the foot of a wall due to the thrust of the untied rafters (as previously indicated in Fig. 9). If the horizontal thrust equals "H" then the moment equals — H x h and the base of the wall has somehow to provide a moment equal and opposite to this. Anti-clockwise moments are called negative and clockwise moments positive.

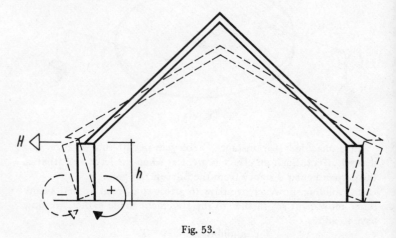

Fig. 53.

A situation in which forces are guided along rafters but do not ultimately converge on one point is illustrated in Fig. 54. The traditional collar beam roof is an everyday example.

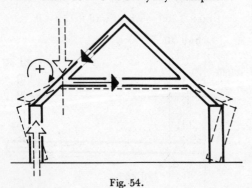

Fig. 54.

Bending Moments

The most common of all structural detours is presented by the *beam* in all its ramifications (Figs 55, 56). The further a load is removed

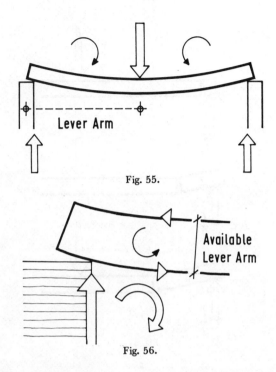

Fig. 55.

Available
Lever Arm

Fig. 56.

from the supports, the greater will be the moment acting on a beam. In other words, the longer the beam, the greater the moments it has to withstand. To be able to supply the necessary equilibrating moments without bending excessively or breaking, the beam has to provide forces within its own thickness. As the thickness is necessarily small in relation to the span, the available lever arm is small and the "beam forces" produced will have to be considerable. It is therefore important to be able to appreciate the distribution of moments along beams and, in fact, in any structural members that are subject to bending. The name given to these moments is *bending moments*.

Basic beam action

Figure 57a, b, c shows the left hand end of such a beam. If it is loaded by a force P in its centre, the support "pushes up" at the rate of $P/2$. At a distance 1 unit of length from the support the moment equals $P/2 \times 1$. At 2 units distant, it equals $P/2 \times 2$; at 3 units away from the support, $P/2 \times 3$ and so on. The support at the other end produces the same moments but these rotate in the opposite direction (Fig. 58).

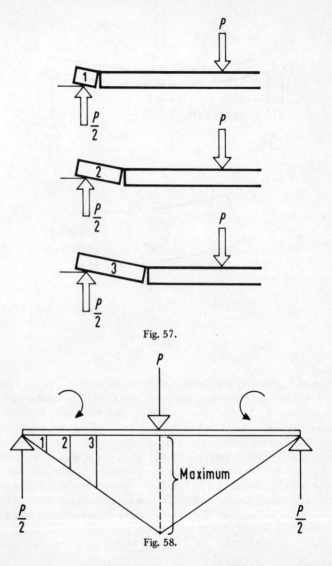

Fig. 57.

Fig. 58.

These moments can be plotted and the resulting graph indicates the distribution of moments along the beam, with a maximum, in this case, occurring at the centre. Already, it is apparent that a beam must be designed to withstand this maximum, at least in the centre. The concept of *maximum bending moment* is therefore of vital importance. When there are two symmetrical loads the moments are distributed differently (Fig. 59). In this case, the maximum is spread over a considerable distance along the span.

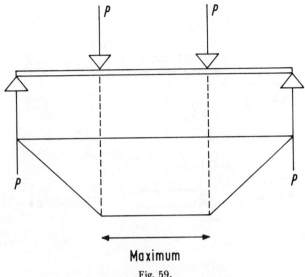

Maximum

Fig. 59.

A numerical example is shown in Fig. 60a, b, c. Let the load be 50 kg, acting 3 units from the support which also provides a reaction of 50 kg. At the load point, as before, the moment has reached a magnitude of 50 x 3. At a point 1 unit further along, the load (if left to itself) would turn the beam back at the rate of − 50 x 1, so here we have

$$+ 50 \times 4 - 50 \times 1 = 50 \times 3$$

At a point 2 units along, we have

$$+ 50 \times 5 - 50 \times 2 = 50 \times 3$$

and so on.

The bending moment along the balcony shown in Fig. 48 grows linearwise, from zero at the "free" end to a maximum at the supporting end (Fig. 61). To represent bending moments in the form of clearly intelligible graphs, those which produce curves in one direction (in the case of beams downward) are referred to as positive. If producing opposite and upward curves, they are negative. Figure 62a indicates a beam which over its entire length would curve *upwards*. Its bending moment graph would thus be entirely negative (Fig. 62b). In Fig. 63a is shown a beam with one end cantilevered. Between supports the graph would curve *downwards* (positive) and at the cantilevered part *upwards* (negative). Somewhere along the span is a point at which it would curve neither upwards nor downwards. This is called the *point of contraflexure* where in

33

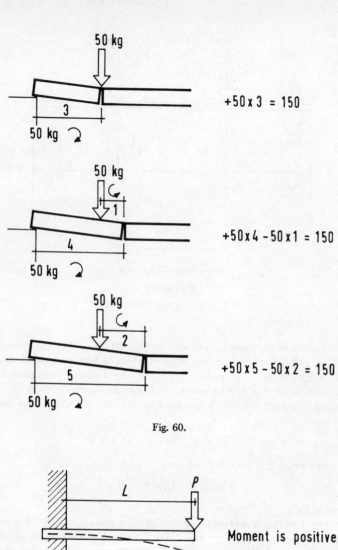

+50 x 3 = 150

+50 x 4 - 50 x 1 = 150

+50 x 5 - 50 x 2 = 150

Fig. 60.

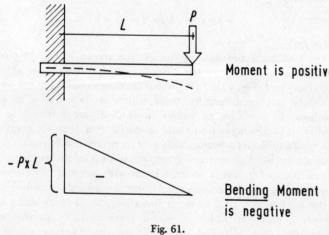

Moment is positive

Bending Moment
is negative

Fig. 61.

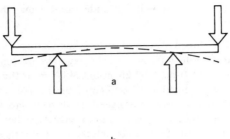

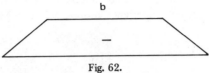

Fig. 62.

theory, the beam would remain straight. At this point the bending moment passes through zero (Fig. 63b). Here, the beam could be quite thin and, in fact, a hinge could be inserted (Fig. 64). The more that loads are distributed along a beam, the smaller are the resulting

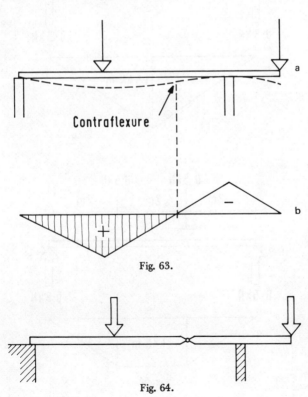

Contraflexure

Fig. 63.

Fig. 64.

bending moments. This will be understood from the following numerical example:

In Fig. 65 is shown a beam 6 m long supporting a load of 1 kN, distributed from 1 to 5 points respectively. The maximum bending moment is reduced from 1.50 kN m to 0.9 kN m. If the load is fully distributed, the graph will become a continuous curve. The result is a further proof that for the same overall load, distributed loads must result in lighter structures. Once again the structural detour becomes shorter as more of the loads approach the equilibrating supports. Of course this is common experience and Fig. 66 should drive the point home!

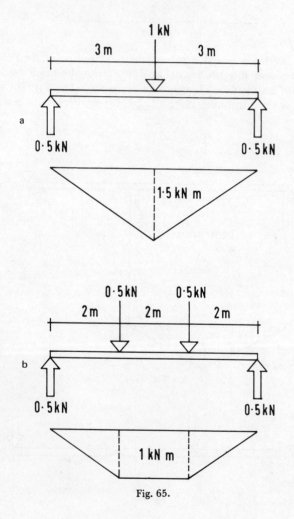

Fig. 65.

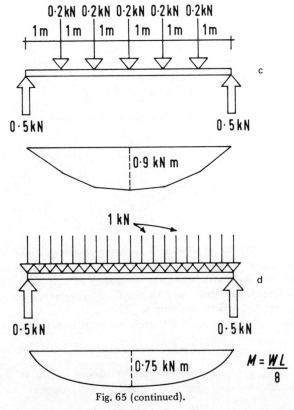

Fig. 65 (continued).

The case of the distributed load is common for all ordinary beams, joists, lintels, etc., whenever a floor or a wall is to be supported.

The numerical relationships for a centrally supported point load, and for a distributed load of equal magnitude is very simple. Because they are so common and also so useful we include them here:

In Fig. 67, the diagram shows a beam Lm long loaded centrally by a load of W kN. The reaction (support) at each end equals $W/2$, so that maximum central bending moment equals

$$M = W/2 \times L/2 = \frac{WL}{4} \text{ kN m}^*$$

Fig. 66.

* Note that live loads are measured in "kN" (kilonewtons) and bending moments in kilonewtons x metres.

37

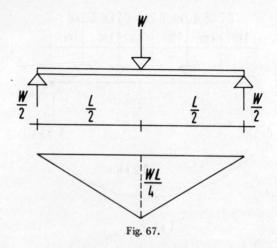

Fig. 67.

In Fig. 68, the same load W is shown as spread evenly over the whole length of the beam. To obtain the bending moment in the centre, the load is divided into two halves of $W/2$ each. To obtain the moment, each half may be replaced by its resultant acting as shown at: $L/4$ m from the centre

The reactions are again $W/2$ each. So the moment at the centre is:

$$M = W/2 \times L/2 - W/2 \times L/4 = \frac{WL}{8}.$$

Ideally, a beam needs to have sufficient depth to meet the maximum bending moment whenever it happens to occur.

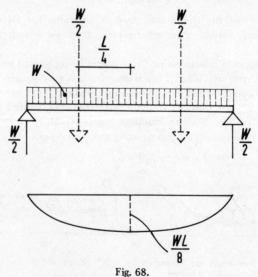

Fig. 68.

38

Generally, however, this is not a very practical proposition. There may be a case for the ideal in roofs, but certainly not in floors! (see Fig. 3).

The foregoing indicates the considerable influence of bending moments on structural members. Detailed calculations can be found in many specialized textbooks, but the ability to visualize bending moment distributions qualitatively and to relate these to the structural forms is of vital importance.

The no-moments line and moments on portal frames

Bending moments have been shown to result from the inability of certain structural members to guide forces to the ground in a direct route. A cable is unable to transmit any bending moments, nor is a chain, due to its many links. Under load, a cable will therefore fall into the line of least resistance. As we have seen in Chapter 2, this is the funicular line, which can also be described as the *no-moments line*.

Any structural form bridging the same span will consequently develop moments which are *in direct proportion* to the deviation of this line from its basic form. We have already seen the beam as the worst offender in this respect. On the other hand the arch (especially when parabolic) comes closest to the no-moment line. Ideally it transmits no bending moments at all. Between these two extremes are numerous structural forms designed to approach the funicular line to a greater or lesser degree. These come under the category of *portal frames*.

A single point line would be represented by a funicular line as shown previously in Fig. 27. The obvious portal frame to carry this load would be a pair of struts (Fig. 69). For two loads, it could be a frame as shown in Fig. 70. The whole frame would be in pure compression and the horizontal thrust at the feet would depend on

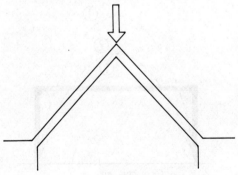

Fig. 69.

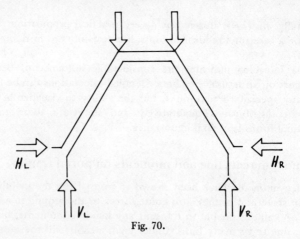

Fig. 70.

the inclination of the legs. Figure 71 illustrates a square portal frame and shows that the no-moment line deviates considerably from this. The greatest moments occur at the "knees" and are proportionate to the distance d. This results in the usual thickening of the frame at this position.

Obviously a great number of no-moment lines could be drawn for such a load arrangement (Fig. 72). The more this line deviates from

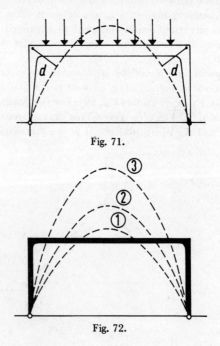

Fig. 71.

Fig. 72.

the beam part of the frame, the larger is the beam moment and the smaller the moment at the knees. Much depends, therefore, on how much the uprights are capable of restraining the beam. This is illustrated diagrammatically in Fig. 73. The restraining ability depends on the relative *stiffness* of the beam and supports. Two

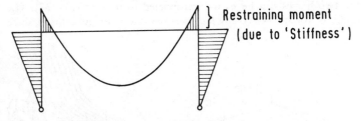

} Restraining moment
(due to 'Stiffness')

Bending moment distribution in
a situation similar to ③ in Fig 72

Fig. 73.

extreme cases are shown merely to indicate what would happen if the restraint were very great; Fig. 74 with the supports taking the moments fully, while Fig. 75a, b shows a situation in which the supports cannot transmit any moment at all and the system then reverts to a simple beam on two supports. The exact dimensioning of these frames involves complex analyses rather beyond the scope of this book, but basic problems involved will be summarized later.

Ideally, moments through the beam part should be as small as possible. To keep moments zero in the beam, a hinge is introduced and as a hinge cannot transmit moments, the no-moment line must

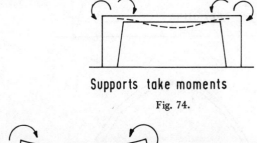

Supports take moments

Fig. 74.

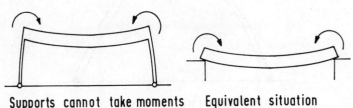

Supports cannot take moments   Equivalent situation

a   Fig. 75.   b

41

pass through it. In such cases there is no doubt about the no-moment line and the arrangement is statically *determinate* (Fig. 76). The resulting structural form is called a *three-hinged* frame. The three-hinged portal is a very simple and frequently used structure. It is easy to erect and for medium-sized buildings, such as farm sheds, the two halves can be easily transported to the site (Fig. 77). The frame is not subject to any local stresses due to irregular loads, since

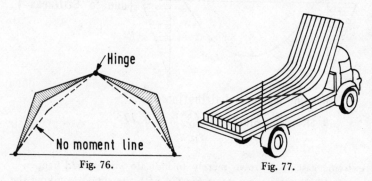

Fig. 76.

Fig. 77.

the three hinges do not transmit any bending moments. However, its practicability is lessened when the span in relation to the height is too great. This will be understood by reference to Fig. 78a and b. In Fig. 78a the no-moment funicular line deviates too much from the ideal form causing very great moments at the knees. On the other hand, the shape indicated in Fig. 78b would be ideal. In most cases,

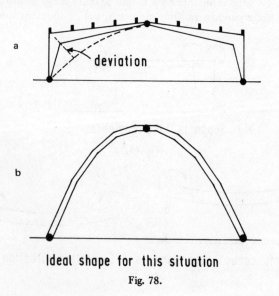

Ideal shape for this situation

Fig. 78.

42

loads are continuous or nearly so, so that the no-moment line approaches the catenary. This case will be used in the next examples.

When spans become longer in relation to height, the bending moments in the beam become rather larger and rigidity becomes increasingly important. The most economical height to span relationship is one to two for the *two-pin frame*. The catenary in this case is one in which the height above the beam is equivalent to a moment of $WL^2/16$ (Fig. 80a, b). This occurs when the I values of the uprights and the horizontals are about the same. (Stiffness of uprights double that of the horizontal, see Fig. 79.)

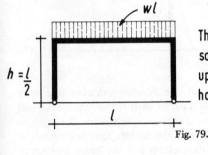

Thickness the same throughout so that the 'stiffness' of the uprights is twice that of the horizontal

Fig. 79.

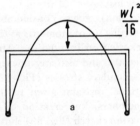

The 'equivalent' catenary gives only an approximation.

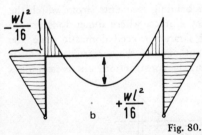

Actual moment distribution shows that the 'beam' moment equals the moments on the uprights justifying the identical cross-section.

Fig. 80.

As these frames are normally used where the rooms are rather high this presumes large spans. One important fact regarding the *two-pin frame* should be noted, namely that the zero-moment or contra-flexure points (see Figs 63 and 64) could be replaced by joints which would act as hinges. This simplifies erection, as two uprights can be erected separately and the centre piece inserted (Fig. 81). The

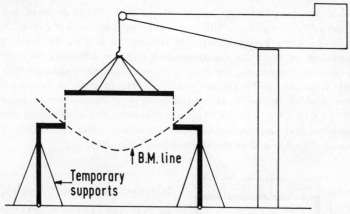

Temporary
supports

↑ B.M. line

Fig. 81.

dotted line in the diagram shows the bending moment line as it would occur, and this may be compared with the catenary. As will be seen later, bending moments increase rapidly with the span — consequently any relative increase in the middle span increases the span moments considerably. This leads to a heavy beam and many advantages of the rigid frame are therefore lost.

The completely rigid frame (*no pin*) gives considerable stiffness to the uprights but due to the setting up of moments at the base of the uprights, eccentricities are exerted on foundations. This makes foundation work complex and costly, and also any unevenness in settlement sets up considerable secondary stresses (Fig. 82). Where frames are very high, the additional restraint given to the uprights will result in more slender members, but erection difficulties can be considerable. The deformation sketch (Fig. 83) shows contraflexure at the change of bending moments, from which the uprights benefit. In multi-storey frames, where dimensions are on the whole moderate and heights rarely exceed domestic scale, the fully rigid frame is a natural outcome of the construction. This is especially so in reinforced concrete construction, but is nowadays

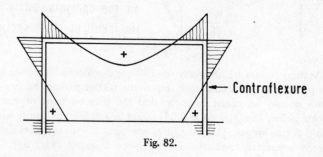

← Contraflexure

Fig. 82.

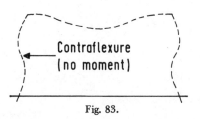

Contraflexure
(no moment)

Fig. 83.

always taken into consideration in steel framing which is welded or connected by high strength friction grip bolts. This results in much lighter frames than in the past.

Calculations for the determination of actual dimensions are not the purpose of this book but, as they represent a major part in the geometry of structural form, a recognition of the statics of the portal frame is essential in the appreciation of structural behaviour.

4 Stresses and bending

Stresses in general

Throughout the preceding chapters it was shown that the forces acting on structural members produce in them reactions which result in equilibrium. As we have seen, these reactions differ in magnitude and intensity according to the structural layout and overall shape of a building.

We can now discuss the nature of these reacting forces inside structural members. We anticipate that they must follow similar patterns and can in fact be subject to analysis based on the same laws of geometry. After all, the fact that the structure of materials seems small to us is merely a fluke. It results from our own terrestrial size which gives us a scale by which, relative to ourselves, we call small things small and large things large. To accept this is to be entirely in keeping with the attitude of modern physics which recognizes the existence of similar laws in the macrocosm of the world as it does in its microcosm.

A force has been defined as a change, producing action. So, if a piece of material is under a compressive force it will shorten (Fig. 84). The molecules will change their relation to each other, but due to their innate inertia, will want to return to their original state. This fact produces the necessary equilibrating force which enables a piece of material to support a load. For instance, it enables us to stand on the ground (Fig. 85). Where this molecular bond is inadequate no equilibrium exists. It is why we cannot stand on a liquid (Fig. 86).

This simple fact was systematically observed by the British scientist Hooke, who stated in 1676: *Ut tensio sic vis,* "like tension, like force", tension meaning deformation. Force has to be qualified as being a unit of force acting on a unit of area. If the force is large in relation to the area affected the tension will be very great (Fig. 87), while a force acting on a large area produces little effect (Fig. 88). This relation of

$$\frac{\text{force}}{\text{area}} \left(\frac{P}{A} \right)$$

is called *stress.* It cannot be defined in any other way, since it is interchangeable with the deformation it causes. In fact, this is the

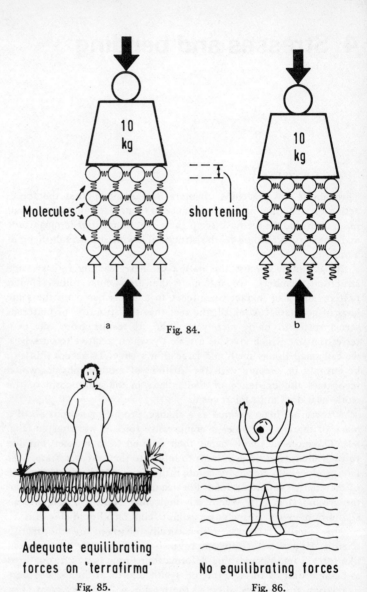

Fig. 84.

Molecules

shortening

a

b

Adequate equilibrating
forces on 'terrafirma'

Fig. 85.

No equilibrating forces

Fig. 86.

only way in which we can become aware of its existence. Deformation is also relative and can only be measured in relation to the size of the unstressed object. This relationship is called *strain*.

In everyday language stress and strain are often used as synonyms, but *stress* defines a relative *load*, and *strain* a relative *deformation*.

48

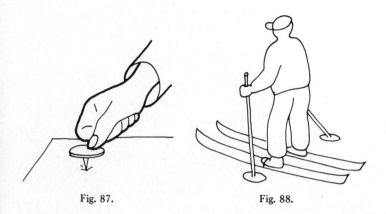

Fig. 87.   Fig. 88.

Elasticity and Hooke's Law

So Hooke's Law can be summarized as

$$\frac{stress}{strain} = \text{a constant.}$$

Now, different materials react differently to stressing. A piece of rubber will deform substantially under a relatively small stress, while steel will require very high stresses before a noticeable strain occurs. More than a century after Hooke, the physicist Young established the rates of this constant for different materials. Identical materials will show identical relative deformations (strains) under identical unit loads (stress). These constants are called Young's Modulus or E for any given material. So we expect the E value for steel to be high and that for rubber to be low. If these ratios are known, we can observe a strain (that is, a deformation), and deduct from this the stress under which it has occurred.

Like so many things we are dealing with in this book, it can be related to everyday experiences. If we stand on a weighing machine, we expect the pointer to tell us our weight. Its movement is in direct proportion to the load which is not measured in terms of relative deformation but in units of weight (Fig. 89). As strain gives a direct indication of stress, this principle was used to test old road bridges for the heavy loads they might have to support under modern conditions. Strain gauges were attached to crucial positions along the arches and as the stress-strain ratio of brick joints was known, the stress could be deduced. As it was known how much stress the arches could safely take, this seemed a very good way of testing bridges without having to break them! All stress analysis is based on this *elastic behaviour* of materials.

Fig. 89.

Stress values

In order to put the foregoing to any practical use we must know the actual stress values involved. We will deal here with the three most important structural materials: steel, concrete and timber.

Steel
When this is tested in either tension or compression, a graph can be drawn (Fig. 90) to show that stress and strain (the rate of deformation) increase at an equal rate, strictly in accordance with Hooke's Law, *up to a point at which the strain increases without an appreciable increase in stress.* This point is called the *yield point*.

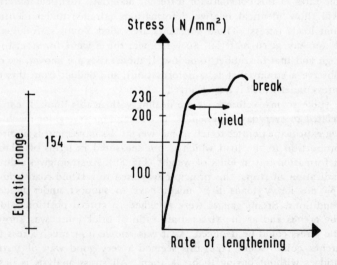

Fig. 90.

After the material yields beyond recovery, the stress will once more increase slightly up to the ultimate stress, after which it will break. Once beyond the yield point the steel will no longer return to its original length, so from a construction point of view, it has lost its usefulness. The phenomenon will once again be familiar to anyone who has ever stretched an elastic band beyond a certain point where it loses its elasticity. So for the time being, we are only concerned with that part of the graph which agrees with Hooke's Law. The yield point occurs at 15 t/in.^{2} or 235 N/mm^2, but as a precautionary measure, we only use two-thirds of this value, giving us a working stress of 10 t/in.^{2} or 155 N/mm^2.*

Concrete
The strength of concrete varies with the proportion of the mix, the aggregate, whether hand or machine mixed or vibrated and also with age. So no precise values can be given for this material. Working stresses are arrived at as a safe fraction of the strength of a test cube. These vary from about 750 lb/in.^{2} or 5 N/mm^2 to about 4000 lb/in.^{2} or about 28 N/mm^2.

Timber
This is a natural product and its strength varies a great deal between different kinds of wood. Within these categories it is subject to grading, a process of selection in which are taken into consideration the grain, number of knots and their distribution, "shakes", moisture content and other criteria. However, common working stresses for timber have been adopted which vary between 800 lb/in.^{2} or about 5.5 N/mm^2 to 1200 lb/in.^{2} or just over 8 N/mm^2.

Young's modulus for steel, concrete and timber

This is just as important as the actual strength values and sometimes more so. There are quite strong materials which have such low *"E"* values that they are all but useless for structural purposes. Other materials possess high *"E"* values but only have a very limited range of strength.

Referring again to the previously mentioned materials:

Steel
13,000 t/in.^{2} or 210,000 N/mm^2.*

Concrete
As it does for strength, this value varies with the mix and the density, etc. from 2,000,000 lb/in.^{2}, or 14,000 N/mm^2 to 3,000,000 lb/in.^{2}, or 21,000 N/mm^2.$^-$

* Values for mild steel.

Timber

Here again considerable variations occur, from about 1,200,000 lb/in.^{2} or 8400 N/mm^2 to 2,000,000 lb/in.^{2} or 14,000 N/mm^2.

It is interesting to note that there exists a strong resemblance in the various values between concrete and timber.

5 The elastic theory of bending and the moment of inertia

We have already seen that bending plays a major part in structural behaviour and that this is caused by bending moments. We have also seen that to equilibrate moments, opposing moments are required. Figure 56 shows how a beam under bending stress has to react internally to produce these equilibrating moments. It was also explained above that deformation is indicative of the stresses involved. These stresses, which are the result of the elastic deformation, produce the forces required for equilibrium. Figure 91 shows a beam under load and its deformation at one section. We see that the greatest *shortening* occurs at the top causing *compressive* stresses, while the greatest *lengthening* occurs at the bottom causing *tensile* stresses.

The deformation in this case is in exact proportion to the distance of individual beam particles from the centre. A particle of fibre at a distance *e* from the centre extends by one unit of length; a fibre at twice that distance extends twice as much; a fibre 3 x *e* from

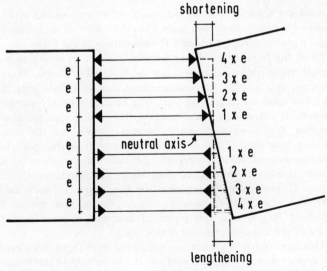

Fig. 91.

the centre extends by 3 units and so on. It can be inferred from Hooke's Law that the stress in each fibre increases at the same rate.

With the centre line as fulcrum, we now have the necessary set of forces to produce the counteracting moment. The total moment is composed of the sum of the individual moments and, as a moment equals force times lever arm, we have a force (acting on a unit of area) times a lever arm of 1, plus *twice* the force (over the identical area) times *twice* the lever arm, plus *three times* the force times *three times* the lever arm and, in our illustration, a force four times as large acting on four times the lever arm, or a total moment about the centre line in the top half of: $1 \times 1 + 2 \times 2 + 3 \times 3 + 4 \times 4$ and the same in the lower half. Figure 92 shows this numerically.

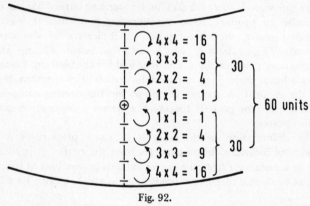

Fig. 92.

Whatever the unit of force, the *combined moment* grows with the *square* of the distance of area units from the centre of the beam. The *actual* moment will depend upon the magnitude of the force.

As all the fibres in our beam are *inert* and try to return to their original length, this moment is called the *moment of inertia*. This is a misnomer as it is only a *potential* moment, one which *would* occur if the beam were actually to be bent. The term is generally accepted, however. It describes a geometric property of a section, because it describes the moment provided by area particles of the section resulting from their distance (hence lever arm) from the centre line. The moment of inertia therefore depends on the *shape* of the section. It must be emphasized that this geometric relationship only applies if two adjacent sections are very close to each other. Otherwise the relative distances would change too much. The important fact is that this potential moment increases with the *square* of the distance from the centre line.

This line, which is assumed as remaining unchanged, is called the *zero line* or *neutral axis* of a section. It can be proved mathematically that the neutral axis passes through the centroid of any section (any line parallel to any edge which passes through the centre of gravity). In symmetrical sections it is, of course, the actual centre line.

Figure 93 shows that, in fact, considerable bending would have to occur before any adjacent sections were substantially deformed. It can be seen from Fig. 94 that the moment of inertia — in future to be referred to as I — grows only *linear* with the *width* of a section.

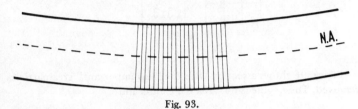

Fig. 93.

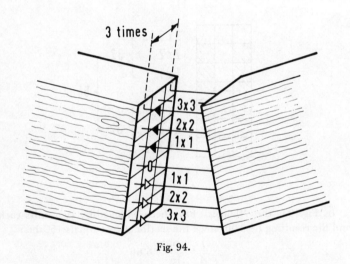

Fig. 94.

The moment of inertia is an indication of the *stiffness* of a section irrespective of the material involved. *Stiffness* has already been referred to in the resistance to bending of members in portal frames and will now be further explained.

If stiffness is a desirable quality in a section we can compare the "efficiency" of different sections in relation to the material used.

Figure 95 shows a section 7 area units deep and 3 units wide. Its I value is 84, its area 21 units, so

$$\frac{I}{A} = \frac{84}{21} = 4$$

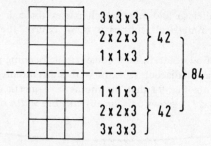

Fig. 95.

Figure 96 shows a section with the "less important" area particles removed. The I value is 64, area is 11 units and

$$\frac{I}{A} = \frac{64}{11} = 5.82.$$

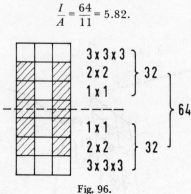

Fig. 96.

In Fig. 97, the "flanges" have been widened by two units each and the resulting $I = 100$; since the sectional area equals 15, then

$$\frac{I}{A} = \frac{100}{15} = 6.66$$

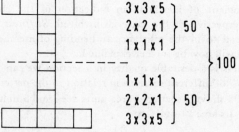

Fig. 97.

This section is 64% more efficient in terms of material than the one shown in Fig. 96. The shape is, of course, well known as the shape of the RSJ, the familiar rolled steel joist. Figure 98 shows a square section of the same depth. In this case

$$I = 146, \quad A = 24, \quad \text{and} \quad \frac{I}{A} = \frac{146}{24} = 6.08$$

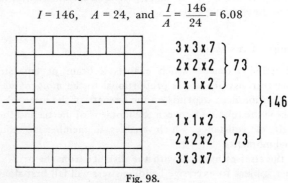

Fig. 98.

Finally, by laying the section shown in Fig. 97 on its side the I value now becomes only 24, area equals 11 so

$$\frac{I}{A} = \frac{24}{11} = 2.17$$

only *one third* of the same section when used upright.

In the foregoing examples for the sake of simplicity the area particles have been assumed as rather large. By adopting very small area units the values will be more correct and slightly larger. The correct values for different basic shapes are given in Fig. 99a-d. The

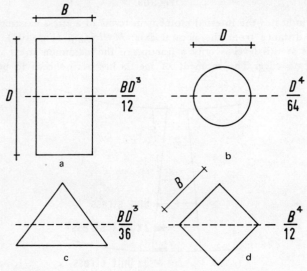

Fig. 99.

mathematical derivation for these can be found in most relevant textbooks. By comparing the correct I values for the 7 x 3 "section" shown in Fig. 96, this will be found to be slightly larger — 85.75 as against 84. It will be noted that in Fig. 99d the moment of inertia is given for the *diagonal* axis.

Moment of resistance

The internal moment which enables a beam or any structural member to resist bending is proportional to the moment of inertia. The *actual* moment depends on the strength of the material. We will now show the relation between the moment of inertia and the actual moment of resistance which enables a member to equilibrate external moments.

As the stresses increase with the distance from the neutral axis, a member subject to excessive bending stress will fail first along either edge, either in compression or tension or a combination of both. This is where the maximum stress occurs (expressed as f_{max}) and the situation is illustrated in Fig. 100.

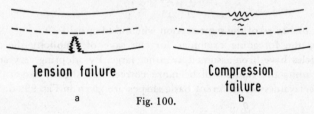

Tension failure   Compression failure

a   Fig. 100.   b

I indicates the internal moment in terms of a stress existing at a unit distance from the neutral axis: $M = I \times f_{unit}$ (Fig. 101). The stress at this level is only a fraction of the maximum stress at the extreme edge. The moment of inertia has been shown to be the

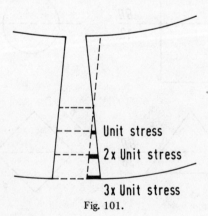

Unit stress

2x Unit stress

3x Unit stress

Fig. 101.

58

summation of all the stresses as they are distributed over the section, and of their lever arms. If the maximum stress is, say, 3 times as great as the unit stress, the above equation will read $M = I \times f_{max} \times \frac{1}{3}$. If the half section above or below the neutral axis is y units deep, the above relation for the internal moment has to be divided by y. So in terms of maximum stress (f_{max}) the expression becomes

$$M_{res.} \text{ (Moment of resistance)} = I \times f_{max} \times \frac{I}{y}$$

But both I and y depend only upon the shape of the section and therefore can always be combined into one, namely I/y. So the final relation between an external bending moment M which has to be equilibrated by the resisting moment M_{res} becomes:

$$M_{ext} = M_{res} = \frac{I}{y} f_{max}$$

This is the most important of all relationships in what is called the *elastic theory of bending*.

In most cases the value which is permitted for the maximum stress is known. So, given a certain shape, it is possible to arrive at the moment of resistance a beam or other member can provide against bending. The relationship I/y, the two parts of which always occur together, is given the name of the *elastic sectional modulus*, symbolized as Z. It expresses in terms of maximum stress the combined forces and their combined lever arm which add up to the moment of resistance.

Elastic theory in practice

The elastic theory can be clarified by giving two examples, the first of which is for a rectangular beam.* The unknown quantity is always the sectional modulus, I/y, and since an infinite number of answers is possible the problem usually resolves itself into either one of trial and error or into making certain assumptions. In the case of timber joists, which are cut to standard breadths, assumptions are necessary. Figure 102 shows the sectional modulus for a rectangle and a numerical example will be found in Appendix 3.

The next example (Fig. 103) explains the method for steel joists. Here, no assumptions can be made, since the sectional modulus depends on the depth and thickness of flanges or webs, and too many variations are possible. Steel joists are therefore tabulated and the sectional modulus required can be found in published tables. Note, however, that the smallest sectional modulus may not always

* See alternative approach to beam design, Chapter 8.

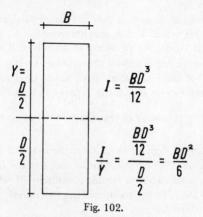

$$Y = \frac{D}{2}$$

$$I = \frac{BD^3}{12}$$

$$\frac{D}{2}$$

$$\frac{I}{Y} = \frac{\frac{BD^3}{12}}{\frac{D}{2}} = \frac{BD^2}{6}$$

Fig. 102.

be the most economical, as deeper sections may sometimes be lighter in spite of higher sectional moduli!

To sum up the procedure for calculation:
1. Determine the maximum bending moment.
2. Insert numerical values into the equation

$$\frac{I}{y} = \frac{M}{f_{max}}$$

3. Find a suitable section with corresponding I/y value, for rectangular timber joists by assuming one dimension, usually the breadth B, for steel joists (RSJs) by the use of tables. For built-up sections in any material a process of trial and error must be adopted.

The above principle applies to any structural members subject to bending provided they are *homogeneous*.* Where different materials are used in conjunction with each other different methods apply, for example, in reinforced concrete (see Chap. 5).

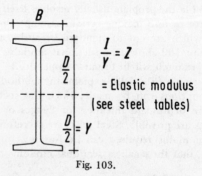

$$\frac{I}{Y} = Z$$

= Elastic modulus
(see steel tables)

$$\frac{D}{2} = Y$$

Fig. 103.

* See alternative approach to beam design, Chapter 8.

Beam theory and beam action

Beam action is the most important principle in the behaviour of structures, and the underlying theory of elastic deformation is the most important aspect of analysis.

Beam theory shows that in any element subject to bending internal moments are created due to the forces not being able to cancel each other out immediately since they are not acting on the same line of action. These moments are responded to by forces of compression on one side and forces of tension on the other. The greater the lever arm between the opposing forces, the greater the resisting moment, which means the deeper a beam the stiffer it is (Fig. 104). It also means that the members which transmit these

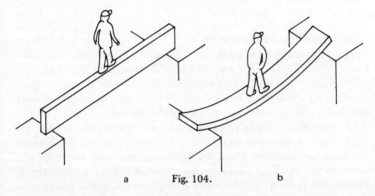

a   Fig. 104.   b

forces can be lighter. This is valid even in suspension structures, in which the tension is taken by a light cable due to the great depth of the "beam", while the compression is taken by the ground (Fig. 105). The resistance to bending in a rectangular section increases

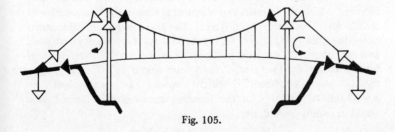

Fig. 105.

with the *square* of the depth but, in common with all sections only *linear* with the width (Fig. 106). It will now be understood how important it is to place the bulk of the material *away* from the neutral axis (see earlier comments on comparative efficiencies).

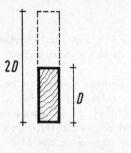
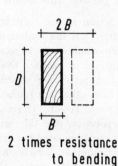

4 times resistance
to bending

2 times resistance
to bending

Fig. 106.

The aim should always be to achieve this with a minimum increase in weight. One method of doing it is to use the *castellated beam*. The castellated beam, which is illustrated in Fig. 107, will take two or three times the bending moment of the original without increase in weight and is invaluable where, due to length of span rather than loading, bending moments become large.

Trusses follow the same pattern of beam action, except that the tensile and compressive stresses are confined within the thickness of the members (see Figs 4 and 5). In view of the small sectional area available in these members, trusses generally have to be deep in relation to their span, but where considerable depth is possible (as for instance in roofs), they form a very light form of beam, again, especially where bending moments are relatively large.

Similarly, for long spans with light loads (as for instance over school halls and class rooms), joists are developed into *trussed joists*. These are usually welded up in light steel angles or tubes with tubular diagonals. Figure 108 shows the beam action. The bending moment which these joists can transmit is simply the compressive *or* tensile force times the depth (lever). The force is the allowable stress times the area ($p = f_{max} \times A$). Figure 109 represents a trussed joist, 500 mm deep. It may be built up of a top and bottom tube of 50 mm diameter. The area of each tube would be 302 mm^2, f_{max} could be 120 N/mm^2 which would be equivalent to $p = 37,000$ N/(37 kN). So the bending moment this trussed joist could transmit would be:

$$M = 37 \times .500 = 18.5 \text{ kN m}$$

The function of the diagonals is mainly to keep the tension and compression zones apart — like the web of an RSJ. They contribute little to the bending strength, but do transmit shear forces. Shear is discussed in Chapter 6.

Fig. 107. The "Castella" beam applied for long spans and light loads

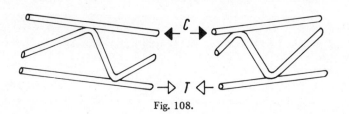

Fig. 108.

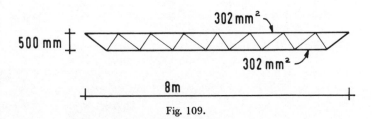

Fig. 109.

6 The reinforced concrete beam

Although the geometry of the influence of the load on a beam is identical irrespective of material there is a fundamental difference in the way that a reinforced concrete beam produces the resistance moment. Within very small limits, a plain concrete beam would behave like any other elastic material. Long before the maximum compressive stress in the concrete were reached (which could be very high), it would fail in tension. So the tension zone is replaced by steel and the tensile contribution of the concrete is discounted. This is made possible because the coefficient of thermal expansion of concrete and steel is almost identical and also because a very good natural grip strength (or "bond") between concrete and steel develops during the setting process.

Until recently, reinforced concrete beams were designed as though, when combined with the steel, they behaved like fully elastic entities. This has, in fact, never been the case. Due to the greatly different E values of steel and concrete, this concept has in the past led to complications and inefficiencies. It has now been largely abandoned.

As in all beam situations, the resisting moments have to be created by internal couples of tensional and compression forces, the compression being provided by the concrete and the tension by the steel. The stress pattern of concrete in compression due to bending is described in Chapter 9 and modern analysis is based upon it. The resisting moment is produced by the resultant of the total compressive force times its lever arm, or the total tensile force times the lever arm. It can be seen from this concept that the lever arm equals three-quarter times the effective depth of either a beam or a slab. "Effective depth" is the depth from the edge of the compression zone to the centre of the steel reinforcement. The calculation procedure is usually based on the relationship of: *steel area* times *steel stress* times *lever arm* being *equal to the bending moment.* In this equation, there are two unknowns, the steel area and the lever arm. By assuming a beam or slab depth, a lever arm is established (¾ x depth) and the steel content can be found. The *Code of Practice for Structural Uses of Concrete* can be used as a guide to a minimum depth, but each case has to be assessed on its merits. (See example for a continuous slab in Appendix 7.) As pointed out in Chapter 9 one of the main changes in the philosophy of structural design lies in the fact that, instead of laying down hard and fast rules, the interpretation of requirements now lies largely in the hands of the designer. Only one criterion is of general validity, namely, that of *stiffness*. This means that the deflection of structural elements, or the elastic deformation of entire structures, must not exceed certain acceptable values. What is acceptable, however, is in itself subject to differing interpretations.

Deflection is of particular significance where spans are long in relation to the load carried (see the great influence of length factor l as discussed in Chapter 8), and in the case of cantilevers.

Due to the difficulties in obtaining a reliable E value for concrete, and the even greater difficulty in assessing the moment of inertia of a reinforced concrete beam because of its composite nature, the minimum depths suggested in the Code are a good starting point. Deflection may be checked in a simplified form by regarding a reinforced concrete beam as homogeneous. By using $BD^3/12$ for I where D is the effective depth, and by using 14,000 N/mm^2 for E, any errors will be on the conservative side. This deflection check has been used in the example in Appendix 7 referred to above.

Slabs, T-beams and other shapes in reinforced concrete

Simple concrete slabs have a small lever arm in relation to their width (Fig. 110) and a great deal of concrete is wasted (Fig. 111). By omitting the useless concrete a substantial lever arm can be

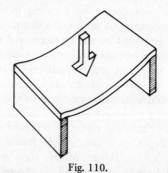

Fig. 110.

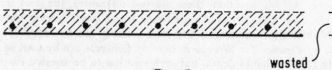

wasted

Fig. 111.

produced without increasing the weight (Fig. 112). This leads to the *ribbed slab*. Sometimes, lightweight tiles are left in position as permanent shuttering when slabs are cast (Fig. 113). Often, removable shuttering made of plywood or plastic is used to produce ribs (Fig. 114). The most economical use is made of the concrete if the neutral axis coincides with the underside of the slab.

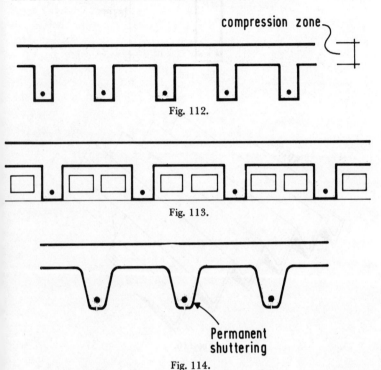

compression zone

Fig. 112.

Fig. 113.

Permanent
shuttering

Fig. 114.

For heavier loading and larger spans, the principle is developed into *T*–beams where the slab portions take the compression (Fig. 115). As there is always an adequate amount of compression concrete available, only the steel has to be checked. The lever arm is taken to the centre of the "slab".

The beam performance of a slab can be considerably improved by folding. *Folded slabs* are a most useful structural form, which is not confined to concrete only. Plywood and plastics can also be used, provided that steps are taken to keep the shape by including adequate cross ties. However, the folds are also subject to beam action from ridge to valley and have to be adequately stiff, that is, fairly thick. This limits the economical depth of folded slabs (Figs 116 and 117). If the stresses can be kept within the "skin", no

67

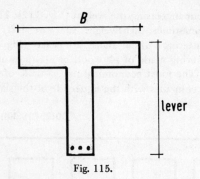

Fig. 115.

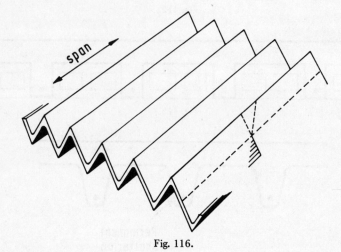

Fig. 116.

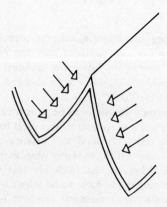

Fig. 117.

transverse beam action occurs. This is the case if the cross-section follows a "funicular", as, for example, in the *concrete shell* shown in Fig. 118. Longitudinally, the shell acts as a *beam*, with its neutral axis approximately in the position shown. Laterally, however, the stresses are confined to the skin, due to funicular arch action. This accounts for the extreme thinness of concrete shells (from 40 to 80 mm). A minimum length to width ratio of about 2 : 1 is needed for efficient beam action, otherwise the shell will act predominantly as an arch (Fig. 119). As with folded slabs, adequate ties are needed to maintain the shape. Again, as in all beams, where bending moments are large due to long spans, depth is needed (Fig. 120).

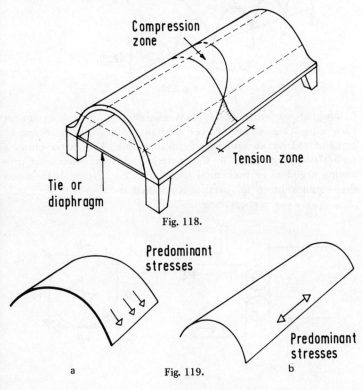

Compression zone

Tension zone

Tie or diaphragm

Fig. 118.

Predominant stresses

Predominant stresses

a   Fig. 119.   b

Shear

When a beam is loaded, some action takes place within it which carries the load to the reactions (supports). We have seen that internal couples are produced in the form of compression and tension, but where do the verticals go? ΣV must be *zero*, and no force can have components at right angles to its line of action.

69

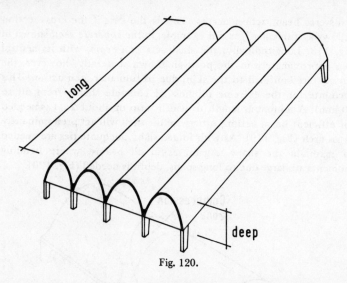

Fig. 120.

Imagine the molecules of two adjacent sections rolling against each other (Fig. 121a). We see a couple of horizontals, T and C, equilibriated by an opposing vertical couple. This latter causes a vertical shearing action in the material, with the horizontal producing a gliding or horizontal shearing action. Figure 121b shows that equilibrium in the particle is restored and also that *horizontal shear must equal vertical shear*.

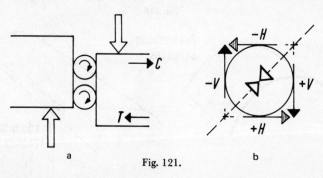

Fig. 121.

The horizontal shear adds up to the compressive and tensile forces along the beam while the verticals are vertically cumulative. *Neither is possible without the other.*

Without adequate resistance to shearing action, a beam cannot take a bending moment. For instance, three planks laid loosely on top of each other would only be as strong as the sum of the three individual planks (Fig. 122), but tied together they would provide a combined beam action 9 times as great (Fig. 123). The increased

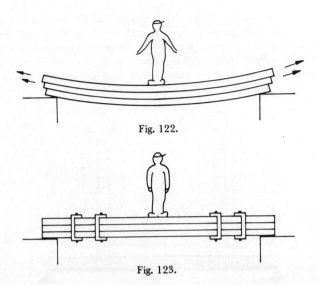

Fig. 122.

Fig. 123.

resistance to bending when horizontal shear is prevented can be shown by holding together the pages of a book, first loosely, and then tightly (Fig. 124). The essence of shear action is that it is the prime factor in the transmission of external forces (loads) to the supports. Longitudinally, it is produced by the difference in sliding action of one area segment over the adjacent one (Fig. 125). This difference remains constant between point loads, resulting in a graph

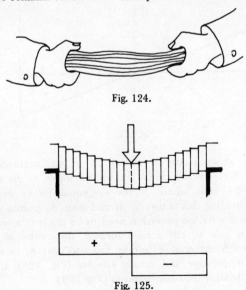

Fig. 124.

Fig. 125.

parallel to the beam. Where the "rotation" of the particles may be regarded as "clockwise", shear is said to be *positive* and where "anti-clockwise" *negative*. Where the change-over occurs lies a zone of "zero" shear (Fig. 126). Where loads are distributed, the differences change and the graph results in a sloping line (Fig. 127a and b).

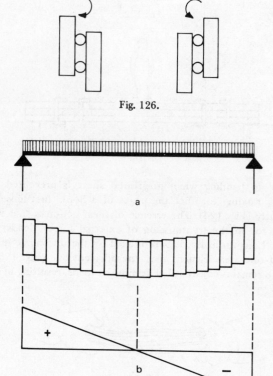

Fig. 126.

Fig. 127.

Maximum shear occurs at the point of maximum difference between the forces acting down and up, obviously at the supports. Shear becomes zero where forces up equal forces down. For symmetrical loading this is usually at mid-span. At points where no sliding action occurs, no provision need to be made to arrest it. For instance, it is safe to cut a hole through the centre of a beam, provided enough material is left at top and bottom to meet the horizontal compressive and tensile stresses (Fig. 128), but holes would be inadvisable near the supports (Fig. 129).

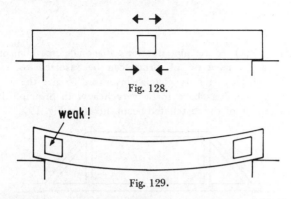

Fig. 128.

weak!

Fig. 129.

In *trusses* the sliding action (i.e. shear) is taken up by the diagonals which act either in tension or compression (Fig. 130). The ends of trusses are often made solid to accommodate a high shear stress in that position (Fig. 131 and also Fig. 8). As with beams the diagonals may be omitted in no-shear positions (Fig. 132). This accounts for the stability of the traditional queen post truss, although part of the inevitable shear is taken by a very thick bottom

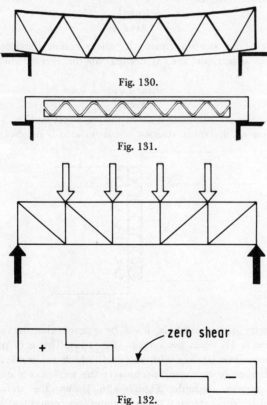

Fig. 130.

Fig. 131.

zero shear

Fig. 132.

73

beam (Fig. 133). In certain situations, diagonals create an obstruction and a special type of girder can be resorted to, called a Vierendeel girder, after its Flemish inventor. To give the necessary stiffness the members have to be very heavy. In principle it is an enlarged version of the castellated beam shown in Fig. 107.

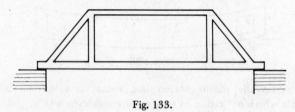

Fig. 133.

Shear stresses and shear distributions

The response of a material to a force acting on it is *stress*, and stress is measured in terms of

$$\frac{\text{Load}}{\text{Area}}$$

To meet the stresses caused by shear action (horizontal and vertical) cross-sectional area is needed. So the average shear stress would be

$$\frac{\text{Shear Load}}{\text{Area}}$$

but as stresses are cumulative down the cross-section (Fig. 134), the distribution of the shear stresses across an area is parabolic. From

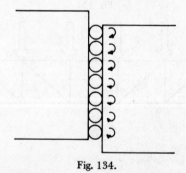

Fig. 134.

the geometry of the parabola, it will be apparent that the maximum shear stress is 1½ times the average (Fig. 135). This is of particular importance when dealing with materials which are weak in shear, such as timber or concrete. In concrete the weakness is due to the lack of tensile strength. Figure 136 shows the deformation attempted by shear stresses. If the compression component is unable

74

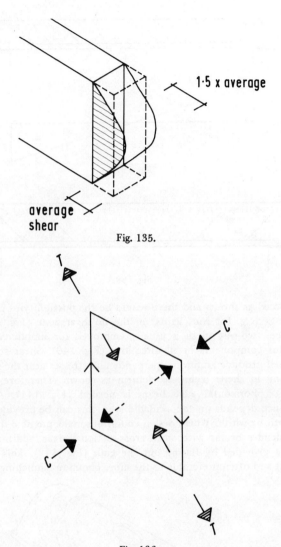

1·5 x average

average
shear

Fig. 135.

T

C

C

T

Fig. 136.

to withstand this, steel bars must be inserted to take up the tension
(which acts at 45°). The number of bars can be decreased as shear
forces decrease along the span (Fig. 137). The arrangement can be
likened to a truss where the reinforcement bars represent the tension
members and the concrete the compression members (see Fig. 138
and also Fig. 130).

What applies in two dimensions can be expanded into three. The
concrete shell acting as a beam (Fig. 118) would ideally be

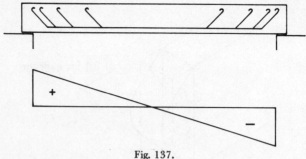

Fig. 137.

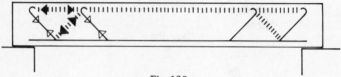

Fig. 138.

reinforced as shown and there would be the possibility of providing openings (e.g. for roof lights) in the no-shear zone (Fig. 139). In practice, two-way mesh is usually employed for simplicity, as this contains components in all directions (Fig. 140). Shear stresses in the shell produce additional perpendicular stresses near the supports resulting in shells trying to deform as shown. Therefore, in long shells, a horizontal edge beam is needed (Fig. 141). As shear resistance depends on the available area, this can be provided either in depth *or* width. Where design conditions make possible the use of very slender beams with small cross-sectional areas, additional area can be provided by flaring out the ends (Fig. 142). This is more elegant and often preferable to the more common haunching.

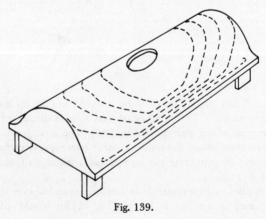

Fig. 139.

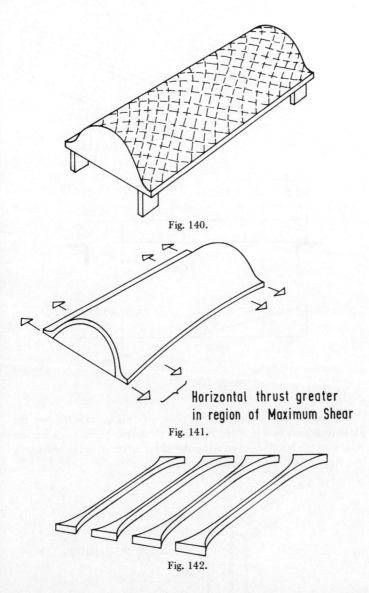

Fig. 140.

Horizontal thrust greater
in region of Maximum Shear

Fig. 141.

Fig. 142.

Due to its grain structure, timber is weak in horizontal shear. In situations of great shear load, a timber beam may split longitudinally (Fig. 143). Box beams and *I*-beams with plywood webs have to be filled in solid at the ends to meet these substantial shear stresses (Fig. 144). The additional area for shear resistance has to be provided up to a point along the span where shear is small enough to be met by the basic beam section itself (Fig. 145).

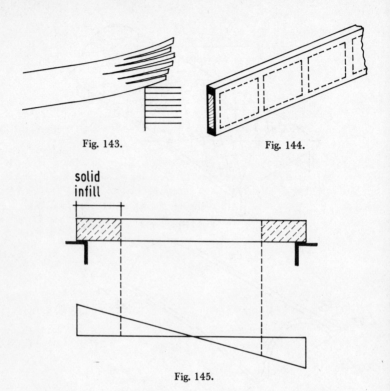

Fig. 143.

Fig. 144.

solid
infill

Fig. 145.

The shear distribution in sections of this nature follows the pattern shown in Fig. 146, but for calculation purposes, adequate values are obtained by averaging the shear stress over the webs only (see Appendix 4 for a detailed example).

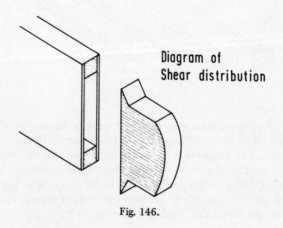

Diagram of
Shear distribution

Fig. 146.

The following table gives permissible shear stresses

	Approximate values		
Steel	108	N/mm²	
Concrete	0.7	N/mm²	(1 : 2 : 4 Concrete)
Timber	0.7-1.4	N/mm²	

As with bending stresses, shear stresses differ in timber for different grades and in concrete for different mixes. The no-moment line discussed in Chapter 5 is also automatically a no-shear line, as the one is not possible without the other. It will also be obvious that no shear action is possible in a catenary.

7 Compression

Compression in general

Where a structural member is in direct compression, equilibrium is direct and stresses are simply

$$\frac{P \,(\text{Load})}{A \,(\text{Area})}$$

as shown in Fig. 147. Unfortunately, this only applies where compression members are short. In longer members, due to inevitable inaccuracies, forces do not act exactly opposite to each other and turning moments are created which may lead to *buckling*

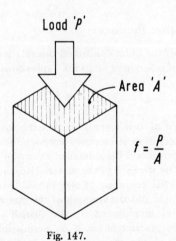

Fig. 147.

(Fig. 148). This tendency increases with increasing slenderness. It inevitably leads to a reduction in the load which a given column can carry. The permissible load therefore depends on the *slenderness ratio* of the column. The slenderness ratio gives an indication of the stiffness of a column, and its tendency to buckle will depend both on this and its elasticity.

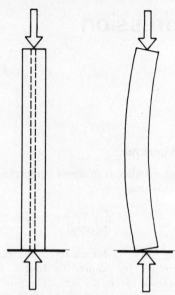

Fig. 148.

Euler and other theories

The Swiss scientist Euler confirmed this relationship for very slender columns and it is summed up in the expression:

$$P = \frac{\pi^2}{l^2} EI$$

The essence of this formula is that the load-bearing capacity of very slender columns is *inversely proportional* to the *square* of the *length* (see Appendix 5 for a detailed example).

Note that the *strength* of the material does not enter into it at all!

For short stiff columns, "Euler values" play a part diminishing with the stiffness and the strength of the material will have a greater share. This was investigated by the British scientist Rankine and others. Present practice is based on permissible stress tables which are the result of the above and experimental observations. These tables refer to slenderness ratios in two ways.

In its simplest form, slenderness ratio is measured in terms of length over minimum thickness. This applies to rectangular sections. For more complex sections (e.g. RSJs or built-up shapes in timber or concrete) the length-over-thickness ratio cannot apply because of the many possible variations. A measurement has to be employed which combines the moment of inertia (as a stiffness factor) with the cross-sectional area, as the moment of inertia alone can apply to a great number of different areas.

In Chapter 5 it was implied that the moment of inertia was the sum of all area particles multiplied by the squares of their distances from the neutral axis:

$$I = \Sigma a y^2 \qquad (\Sigma a = A \text{ (total area)})$$

which leads to

$$y = \sqrt{\frac{I}{A}}$$

Radius of gyration

This means that if the total area (*mass* in physics) could be concentrated into one point at a distance which would give this concentrated area (mass) the same moment of inertia as the original entire section, this distance would be y. Its moment of inertia would then equal Ay^2. This concept sounds very abstract in terms of building, but is well known in dynamics. It can be illustrated on a flywheel. If A represents the mass of the whole flywheel (Fig. 149)

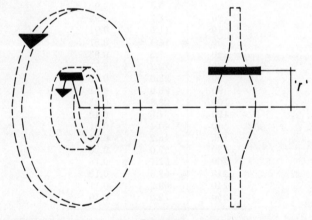

Fig. 149.

this could be replaced by a "concentration" at r millimetres away from the hub having the same moment of inertia which would then be Ar^2 (Fig. 150). As the flywheel "gyrates" the distance is called the *radius of gyration*, r, instead of y. The slenderness ratio can now be defined more accurately as l/r. The adjacent table (Fig. 151) gives reduction factors for the permissible stress for timber in terms of length over radius of gyration and length over thickness.

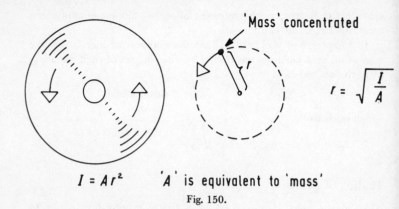

'Mass' concentrated

$r = \sqrt{\dfrac{I}{A}}$

$I = Ar^2$ 'A' is equivalent to 'mass'

Fig. 150.

TIMBER

l/r	l/th	f reduct.
Less than 5	1.4	1.00
5	1.4	0.99
10	2.9	0.98
20	5.8	0.96
30	8.7	0.94
40	11.5	0.91
50	14.4	0.87
60	17.3	0.83
70	20.2	0.77
80	23.0	0.70
90	26.0	0.61
100	28.8	0.53
120	34.6	0.40
140	40.4	0.31
160	46.2	0.24
180	52.0	0.20
200	57.7	0.16
220	63.3	0.13
240	69.2	0.11
250	72.2	0.10

Fig. 151

The table given in Fig. 152 applies to steel columns. As stresses are fixed for steel the table gives the actual reduced stresses in terms of length over radius of gyration. In each case the *minimum* radius of gyration or thickness has to be used. Note the rapid decrease of the allowable stresses with increasing slenderness.

In the case of reinforced concrete the code gives reductions in permissible loads based on effective length (Fig. 153) over minimum

84

STEEL	
l/r	N/mm^2
20	110.2
30	106.8
40	102.2
50	97.5
60	91.0
70	83.7
80	75.5
90	67.0
100	59.0
110	57.5
120	45.3
130	39.9
140	35.2
150	31.2
160	28.0
170	25.0
180	22.0
190	20.0
200	18.7
210	17.0
220	15.6
230	14.4
240	13.3

Fig. 152

CONCRETE	
l/th	P reduct.
15	1.0
18	0.9
21	0.8
24	0.7
27	0.6
30	0.5
33	0.4
36	0.35
39	0.3
42	0.25
45	0.2
48	0.15
51	0.10
54	0.05
57	0

Fig. 153

thickness. The permissible load in concrete columns is shared between the concrete and steel. It is made up, of course, of permissible concrete stress *times* concrete area *plus* permissible steel stress *times* steel area (from $P = f \times A$).

Slenderness ratio

The implication of slenderness ratio versus load-bearing capacity is that long slender columns are essentially uneconomical. This is because their ability to support loads becomes very small in relation to their area, and the material is not used to its fullest advantage (refer to the Euler formula on p. 82). Note also, in the relation for r, this grows with the square root of I and is inversely proportional to the square root of A. So for minimum slenderness I should be kept large and A small. Once again, this implies the advantage of spreading the material away from a neutral axis (see Fig. 98). The table in Fig. 154 shows a comparison between a standard steel section and a hollow steel tube of almost identical cross-sectional areas.

85

Section	Dimension mm	Area 'A' mm²	r min.	Slenderness ratio l/r	f pm N/mm²	P = f x A
RSJ I	155 × 127	4750	28.2	$\frac{3650}{28.5} = 129$	39.8	1890 kg
Standard square steel tube	177.8 × 177.8	4850	69.6	$\frac{3650}{28.2} = 52.5$	95.8	4650 kg

Fig. 154. Performance comparison between RSJ and square steel tube.

This indicates that even allowing for the slightly larger area of the tube, its load bearing capacity is 2.4 times as great. It also shows that an even distribution of the material all round is important as the *critical r* value is the smallest and most shapes and sections have two different values about either axis which usually differ substantially (Fig. 155).

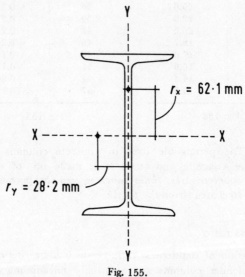

$r_x = 62 \cdot 1$ mm

$r_y = 28 \cdot 2$ mm

Fig. 155.

Oblong sections with greatly differing r values lose rapidly in load bearing capacity with increasing height. The diagram in Fig. 156 shows the relative behaviour of two sections, the r values of which differ to a considerable degree. Their load-bearing capacity becomes identical in the higher regions of slenderness when Euler values are approached.

The tendency of a column to buckle is also affected by the way it is fixed or restrained. The length which will actually initiate buckling is called the *effective length* Figure 157a, b, c shows how the

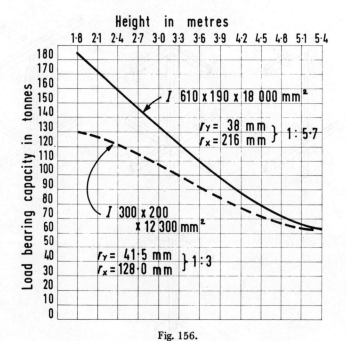

Fig. 156.

Fig. 157.

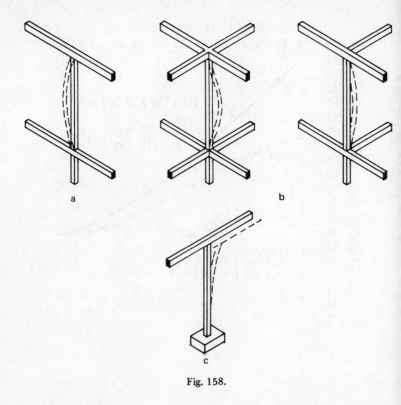

Fig. 158.

effective length l is related to the actual length L and in Fig. 158a, b, c these conditions are related to actual situations in framed structures.

Effective length or height also affects the loadbearing capacity of *walls*. Their thickness is the *actual* thickness with the exception of the "eleven inch cavity wall", whose effective thickness is only 6 in. or 152 mm. In Fig. 159a to e are shown effective heights of walls depending on their connections to floors. Concrete floors usually provide a firm hold, but to act as effective restraints, timber floors have to be fixed to supporting walls with special cramps. The maximum effective slenderness of a wall can be 24. This allows a 105 mm wall (formerly 4½ in.) to be 3250 mm, namely 4/3 times 2520 mm, if properly restrained. This is well over the normal storey height. As for columns, allowable stresses are reduced as slenderness increases. The relevant factors are shown in the table, Fig. 160.

The strength of blocks used for loadbearing walls varies considerably, ranging from lightweight concrete blocks with a permissible strength of about $3N/mm^2$ to engineering bricks more

88

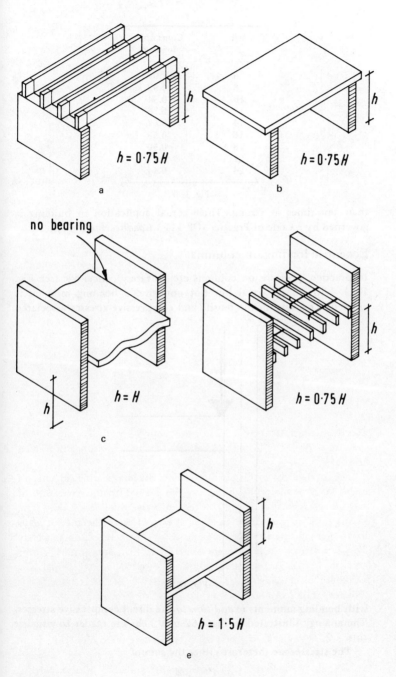

no bearing

$h = 0.75H$

$h = 0.75H$

$h = H$

$h = 0.75H$

$h = 1.5H$

a

b

c

e

Fig. 159.

h/t	Concentric loading
6	1.0
8	0.92
10	0.84
12	0.76
14	0.67
16	0.58
18	0.50
21	0.47
24	0.44

Fig. 160

than ten times as strong. Their actual application in building is governed by a Code of Practice (CP 111 : *Loadbearing walls*).

Eccentric loading on columns

In practice, pressure on columns etc., is rarely concentric (see Fig. 161). With forces equal but not opposite, a bending moment is created. This sets up the tensile and compressive stresses associated

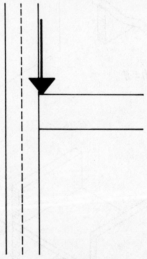

Fig. 161.

with bending moments *in addition to* the direct compressive stresses. The analogy illustrated in Fig. 162 may help the reader to visualize this.

The stresses are therefore either the sum of

$$\frac{P}{A} \text{ and } \frac{M}{Z}$$

Fig. 162.

on the compression side, or their difference on the tension side (refer to Chapter 5). The magnitude of the moment depends on its lever arm, which is the distance of the pressure centre from the centre of the column e. So we have:

$$f = \frac{P}{A} \pm \frac{Pe}{Z}$$

The result is illustrated in Fig. 163a, b, where the two stress patterns

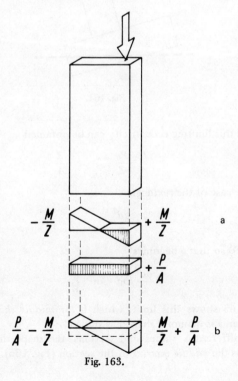

Fig. 163.

are overlaid. There is a tension "residue" which should be avoided with materials weak in tension such as brickwork or unreinforced concrete.

In the case of *foundations* these could be *lifted* off the ground on the tension side (Fig. 164). To avoid such an occurrence, the eccentricity should be limited to a point where *no tension* occurs, in other words, where

$$\frac{P}{A} = \frac{Pe}{z}$$

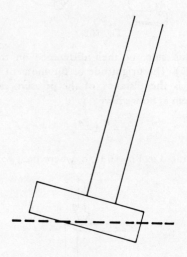

Fig. 164.

From this limiting eccentricity can be obtained

$$\frac{P}{A} = \frac{Pe}{Z} \quad \therefore \quad e = \frac{Z}{A}$$

In the case of the *rectangle*

$$Z = \frac{bd^2}{6}$$

(see p. 59) so that *e* becomes

$$\frac{bd^2}{6} / bd \quad \text{or} \quad \frac{d}{6}$$

Figure 165 shows this limit, which is *the middle third*. The actual limiting area is diamond shaped as shown.

For different *z* values and sections, the shape differs. For the *circle* it is the *middle quarter* of the section (Fig. 166).

92

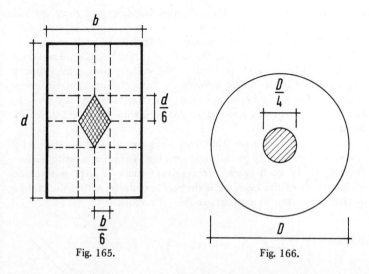

Fig. 165.   Fig. 166.

For hollow shapes the area decreases far more rapidly than the moment of inertia, so that the relation Z/A becomes larger. For thin walled tubes or squares, the limit of eccentricity may become nearly twice as great as for a solid section of the same overall size of diameter (Fig. 167 and see Appendix 6). Once again, it shows the importance of placing material away from a centre.

The limiting area discussed above is called the *kern* of a section from the German word for kernel or core.

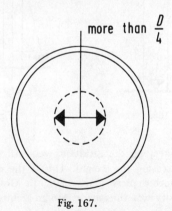

Fig. 167.

Further aspects of the theory of eccentric pressure

Permissible stresses in columns must be kept within the limits resulting from the combination of direct and bending stresses. There are certain exceptions to this in framed buildings and these are laid down in various Codes of Practice.

One aspect of pressure concerns the traditional arch or vault. It will now be understood that the natural thrust line should not only be kept inside the arch but within the *middle third*. This happens automatically if the shape follows the funicular line. In reinforced concrete, this is less important since the steel takes up any tension, but it is still of great importance for very thin shells and generally for reasons of economy.

Eccentric pressure on walls and piers can also be produced by *lateral thrust,* that is, wind pressure (Fig. 168), or slanting rafters (Fig. 169). In such cases, the *resultant* must be kept within the middle third of the base, or, if the pier is circular, within the middle quarter, or within its appropriate limit if hollow.

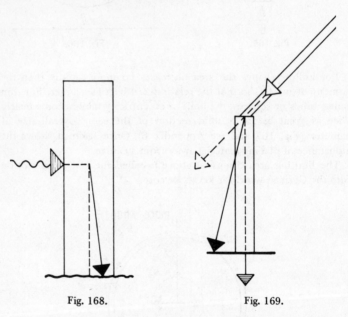

Fig. 168.   Fig. 169.

One way of keeping the resultant well within the limits can be achieved by increasing the weight. This is the main reason for the pinnacles on Gothic piers (Fig. 170). In Gothic architecture a structural necessity was thus made into an architectural feature. The principle applies also to entire buildings. By keeping the resultant within the no-tension limit, no tension occurs on the windward side. This can be of great importance in very tall slender buildings, as it reduces sway and improves weather resistance by avoiding the formation of tensile cracks (Fig. 171).

94

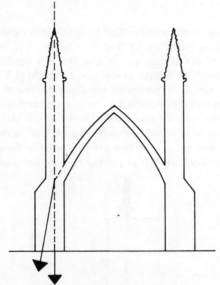

Fig. 170.

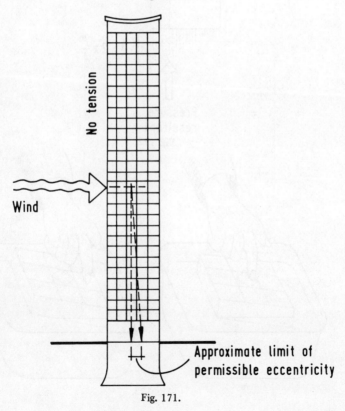

No tension

Wind

Approximate limit of
permissible eccentricity

Fig. 171.

The principle of the middle third complies with the conditions of static equilibrium (see Figs 11 and 23). In Fig. 172 is shown the redistribution of the average pressure so that the equilibriant passes through the centre of gravity of the stress triangle ($\Sigma V = 0$). Figure 173 shows a simple experiment to illustrate the theory. Fill a shallow basin with clay or plasticine and place a rectangular block on top. Press on the edge of the middle third and the block sinks into the base as shown. One edge should remain level with the surface *however hard the pressure*. When pressing *outside* the middle third (Fig. 173b) the block will lift out of the base on the tension side.

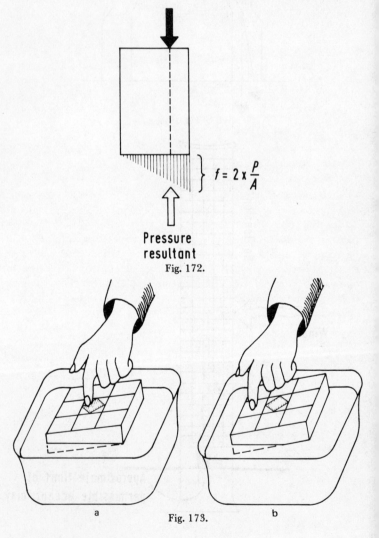

$$f = 2 \times \frac{P}{A}$$

Pressure
resultant

Fig. 172.

a   b

Fig. 173.

Pre-stressing

This is another aspect of the principles discussed so far. In Chapter 6, it was shown that, for design purposes, the tension part of a concrete beam is neglected while the compressional qualities of concrete are not altogether used to their best advantage.

Pre-stressing implies the application of pressure prior to the final loading of a structural member. This creates a pressure "reserve" which can eliminate or greatly reduce tensional stress. Figure 174

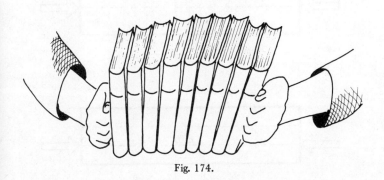

Fig. 174.

shows a well known example. By pressing the lower half, the tensile, and a great part of the shear stresses are eliminated, and the stack of books can act as a beam. Similarly, by tightening up the reinforcement in the tensile zone of a reinforced concrete beam, tension will be reduced and the beam will be able to sustain a much larger external bending moment. The built-in compression creates an opposite bending moment. This is illustrated in Fig. 175, showing the ordinary bending stresses, the superimposed compressive stresses and the stress distribution resulting from an overlay of the two.

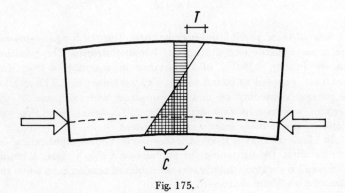

Fig. 175.

The actual stress built up in a beam depends upon the degree of eccentricity of the pre-stressing component. Figure 176 shows three possibilities. In *a* is shown a concentric force resulting in an evenly distributed P/A; *b* shows the pre-stressing force acting on the middle third, producing $2 \times P/A$ on one side and zero on the other; *c* shows an eccentricity of one-third, with a stress of $3 \times P/A$ on the compression side and tension of $-P/A$ on the other side.

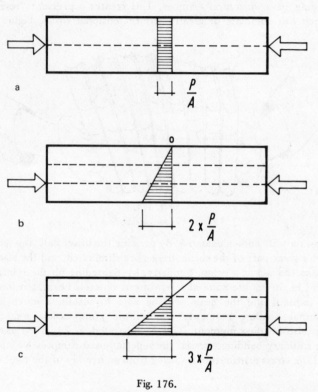

Fig. 176.

Any of these possibilities may be used. The resulting stresses will be superimposed over the normal bending stresses. The situation shown in Fig. 176a is of importance in the case of long slim columns, exposed to lateral forces, e.g. tall lamp posts. The result is a general tightening up and the pre-stress will take care of the tension wherever it may occur (Fig. 177). Apart from this, the concrete is also made more weather-resistant.

In Fig. 178 is shown another well known application of pre-stressing. By tightening the shrouds of a ship's mast, a tensile "reserve" is created which leaves the cables in tension even when the distance is slightly shortened.

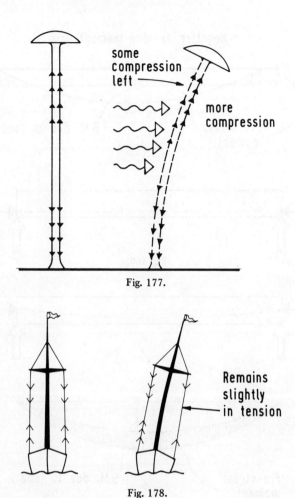

some
compression
left

more
compression

Fig. 177.

Remains
slightly
in tension

Fig. 178.

The aim of all pre-stressing is to reduce the initial bending moments created by a structural member's own dead weight. In many cases, these moments take the shape as shown in Fig. 179. The pre-stressing force P (Fig. 180), produces an even bending moment over the entire beam and if superimposed on the moment shown in Fig. 179 will leave a residual negative moment against which the beam must be reinforced. This can be avoided by sloping up the pre-stressing cables so as to reduce their lever arm gradually (Fig. 181). This results in a moment and stress distribution as shown at b. The slight slope can also take up some of the diagonal shear forces (see Fig. 136). This arrangement further reduces the initial upward curve, or camber, which often occurs in pre-stressed beams. As

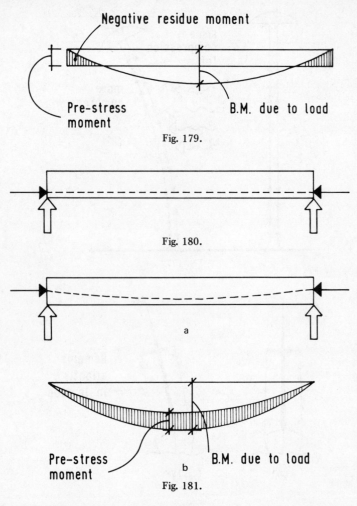

Fig. 179.

Fig. 180.

a

b

Fig. 181.

bending moments are reduced, shear stresses in pre-stressed beams are also accordingly smaller (see Chapter 6).

As concrete gives under pressure and also shrinks on setting, little pre-stress effect would be left after a while, unless the pre-stressing cables were stretched to such an extent that even after shrinkage, considerable tension were retained (see the analogy in Fig. 178). As the E value is the same for steel of any kind, it is important to use high tensile steel in order to be able to extend it considerably. The steel actually used in pre-stressing has 5 to 6 times the tensile strength of ordinary mild steel.

Pre-tensioned and pre-stressed concrete is frequently used in precast beams and planks. Because such components are often

100

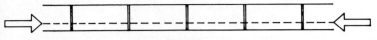

Fig. 182.

manufactured on long continuous bands (Fig. 182) it is not possible
to curve the cables upwards, but shear forces on precast floor beams
are usually small. However, there is often a noticeable upward
curvature which has to be accounted for in screeding. The subject is
covered by a relevant Code of Practice (CP 115).

8 The span–load relationship

The whole of beam theory is based on the deformation which results in curving due to the shortening of the "fibres" on the compression side and their lengthening on the tension side (Fig. 183). If this were constant over the length of a beam, the resulting curvature would be the segment of a circle, but as bending moments change, the stresses change with them and hence also the nature of the curvature (Fig. 184). The curvature changes from zero at the points of no bending

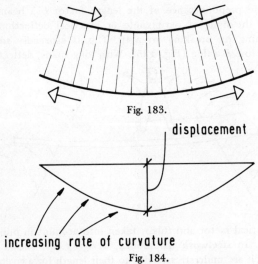

Fig. 183.

displacement

increasing rate of curvature
Fig. 184.

moment to a maximum. A complex geometrical relationship is created between the rate of curvature and actual vertical displacement of a beam from the horizontal. The mathematically-minded reader will realize that the summation of changing values involves the use of the integral calculus and the relationship becomes very complicated due to different conditions of loading.

Deflection

This curvature under load is, of course, deflection. Deflection must not exceed certain maxima. Excessive deflection can be unsightly and can lead to such defects as cracking of plaster, etc. A floor which is allowed to deflect too much would feel rather springy. Codes of Practice lay down a certain maximum value for many simple situations (i.e. $\frac{1}{325}$ of the span*).

Due to Hooke's Law, deflection must be related directly to the load w and be inversely proportional to Young's Modulus E. The shape of the section enters into problems of deflection through the moment of inertia. The larger this is, the less the deflection. Further, for a given load, deflection increases with the third power of the span L, and with the 4th power of the span if the load is given in w per units of length; that is, if the load grows with the length.

So for all deflection

$$d = \frac{WL^3}{EI} \times c \quad \text{or} \quad \frac{wL^4}{EI} \times c$$

where c is a coefficient depending on the type of loading and fixing conditions.

Due to the preponderance of the length factor L a beam may often reach the limit of permissible or practical deflection long before reaching its breaking strength. This is especially so with materials of low E value (Fig. 185). With long spans, deflection is

Fig. 185.

often the critical factor and this is taken into account in published beam tables. In steelwork handbooks, a zigzag line shows beam sections which are understressed due to their length for a given load. These situations should be avoided.

Summary of the span-load relationship based on bending stresses

All structural behaviour is dictated by the essential principles discussed so far. Where stated in mathematical form, as formulae, this merely represents a shorthand description of structural happenings, more of which are discussed below.

* With certain modifications arising out of the load factor approach to reinforced concrete design.

104

Structural form is influenced in the first place by layout, that is, by the overall geometry of a building. This basically governs the shape of the structural elements and so determines their selection and size.

The overall geometry of a building is mainly a question of the lay-out of the members bridging space, and their supports. Ideally, loads should be brought down to earth in the most direct way, but this is rarely possible. Point loads result in greater detours of the forces than those arising from distributed loads (see Chapter 3). This fact is reflected in the bending moment values for a centrally placed point load $WL/4$ and for the same load distributed $WL/8$. In each case for a total load W, bending moments increase with the span L.

For a distributed load w which increases with the span

$$M = \frac{wL^2}{c*}$$

The influence of this relationship on the material needed varies with different *shapes*.

(a) *The rectangular beam*. From $M = fZ$ (Z being I/y):

$$M = f \times \frac{BD^2}{6} \quad \text{or} \quad \frac{WL}{c*} = f \times \frac{BD^2}{6} \quad \text{so:}$$

$$D \cong \sqrt{L} \quad \text{and} \quad \sqrt{W}$$

e.g. double the length and the depth D will increase by $\sqrt{2}$.

If

$$M = \frac{wL^2}{c*} \quad \text{then} \quad \frac{wL^2}{c*} = f\frac{BD^2}{6} \quad \text{and} \quad D \cong L$$

In other words, *depth grows directly with the span*.

(b) *The "open" section*. (Trusses or trussed joists, see Chapter 5.)

$$M = A \times D \ (16.1) \quad \text{and} \quad I = \frac{AD^2}{2}\dagger$$

$$\frac{WL}{c*} = A \times D \quad \text{and} \quad D \cong L$$

* c is a coefficient which depends on the method of support. It equals '8' in the case of the simply supported beam.

† The exact 'I' value equals $A \times d^2 + I_A$ for each half. This becomes

$$2(Ad^2 + I_A) = \frac{AD^2}{2} + I_A + I_A,$$

I_A, the moment of inertia for the cross sections, is usually extremely small and can be neglected in calculations.

(see Fig. 186). In other words *depth grows directly with the span*.

For a distributed load

$$\frac{wL^2}{c^*} = AD \qquad D \cong L^2$$

therefore, *depth grows with the square of the span*.

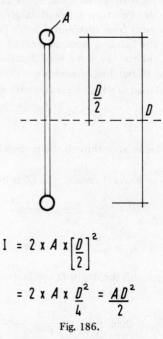

$$I = 2 \times A \times \left[\frac{D}{2}\right]^2$$

$$= 2 \times A \times \frac{D^2}{4} = \frac{AD^2}{2}$$

Fig. 186.

(c) *I sections or similar*. These come about halfway between the solid rectangle and the open section. Generally, it can be said that for a given W the weight of a section increases somewhere between $\sqrt{L}$ and L.

For a distributed load w per unit length, the *selfweight* increases faster than L but does not reach L^2, as in the case of the completely open section.

Span-load relationship based on deflection

For a given material, deflection is governed by the span L and the moment of inertia I. For a given W, the deflection is

$$d = \frac{WL^3}{EI} c$$

and for a distributed load w

$$d = \frac{wL^4}{EI} c$$

(see p. 61). In each case, the deflection depends on L and I.

(a) *For the rectangular beam and a load of W*

$$I = \frac{BD^3}{12}$$

and $D \cong \sqrt{L}$.

So, if D is expressed in terms of L:

Deflection: $\quad d \cong \dfrac{L^3}{\sqrt{L^3}}$

or $\quad d \cong L\sqrt{L}$

therefore, *deflection grows as the span* multiplied by the *square root* of the *span*, that is, *faster* than the *span*. Where $D \cong L$ (continuous load) then

$$d \cong \frac{L^4}{L^3} \quad \text{or} \quad L$$

therefore, *deflection grows with the span*, i.e. relative deflection remains *constant*.

(b) *For the open section*

$$I = \frac{AD^2}{2}$$

for a given W

$$d \cong \frac{L^3}{L^2} \quad \text{or} \quad L$$

therefore, *deflection grows with the span* as above. For w (continuous load) then

$$d \cong \frac{L^4}{AD^2/2}, \quad D \cong L^2$$

hence

$$d \cong \frac{L^4}{L^4} = 1$$

so that *deflection remains the same* (that is relative deflection *decreases* with increase in span!)

(c) *For I or similar sections.* As for bending stresses, deflection comes about halfway between the solid and the open section. As before, however, it is not so much a matter of depth, but of weight.

107

Conclusions on the span–load relationship

(1) The depth of a beam increases only with the square root of the load. So, heavier loading for a given span results only in small increases in depth (and therefore weight of material).

(2) For a constant point or distributed load W, the depth of a beam will only increase with the square root of the span. Therefore, increase in span does not result in a directly proportional increase in material, but deflection grows more rapidly. Therefore, long spans with relatively light loads will tend to deflect excessively *before* maximum stress is reached.

(3) For continuous loads, the *relative* deflection will remain constant but actual deflection will increase due to the increasing selfweight of the beam. This is an important consideration in the case of heavy materials such as steel and concrete. If w is kept small a situation similar to (2) above arises. This is the case with narrowly spaced joists and is also one reason why concrete slabs are uneconomical over long spans (see p. 000 and p. 000).

(4) Where spans are long in relation to loads (e.g. in lightweight roofs), deflection has to be reduced by increasing I without a substantial increase in weight. This leads to "open" sections. Suitably designed, these will rarely deflect excessively as their deflection relative to the span may in fact *decrease*.* Trusses and trussed rafters are therefore ideal bridging elements, provided the necessary depth is available (see Fig. 187).

Small 'w' due to narrow spacing (s)

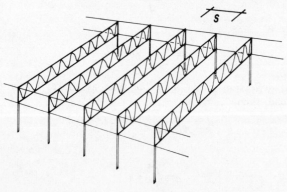

Fig. 187.

* Nevertheless, on long spans, allowance has to be made for deflection for visual reasons and in the case of flat trusses for reasons of drainage. This applies particularly to horizontal space frames.

(5) Conventional steel joist sections are heavy, but if used in the expanded form of "castellated" beams they approach the characteristics of open sections (see Fig. 107). Due to their large I in relation to weight, such beams deflect relatively little. Timber "box" beams etc., have similar characteristics.

(6) Changes in breadth B are discounted, since an increase in moments of inertia and resistance moments derived from breadth are too uneconomical to be considered (see Fig. 106).

To sum up:
Long spans → heavy loads.

Light loads on short spans unless depth can be increased without substantial increase in material.

In other words, where spans are long and loads are light, any kind of lightweight or open section is most suitable, provided that the *depth is available.*

Once again the fundamental economy of *spread* of material is shown.

Shear and span

With depth growing broadly with span in all solid or near solid sections, the necessary cross-sectional area to accommodate shear stresses is normally available. Exceptions occur in the more open sections (see p. 000), where solid end panels may have to be provided.

In the case of short spans, heavy loads can produce excessive shear stresses as bending moments (and hence depth) may be small. Remember that depth grows only with the square root of the load (p. 000).

Precast beams are a typical case of relatively long spans and light loads as the w per unit of length is very small. Therefore, there is never a problem of shear. This is fortunate because, due to the process of manufacture, it would be both difficult and uneconomical to introduce shear reinforcement (see also pre-stressing, p. 000).

Concrete slabs can be likened to a series of closely spaced beams, therefore shear reinforcement is rarely necessary. The same applies to hollow tile floors and those utilizing closely spaced T-beams (Figs 111-115).

Columns

As we have shown, the load bearing capacity of columns is chiefly governed by the slenderness ratio. As this ratio rapidly decreases with an increase in length, long columns are uneconomical. Long in this connection means long in relation to the load. A heavy load requires a large cross-sectional area which makes it a *relatively* "short" column. For its given cross-sectional area such a column will carry a great deal more load. An example of long columns is shown in Fig. 188. Due to the close spacing, each of these will carry relatively little load. Figure 189 illustrates a system of heavily loaded

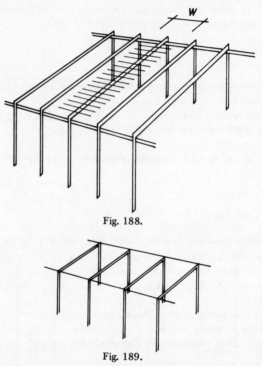

Fig. 188.

Fig. 189.

columns; by using only half the number shown in Fig. 188, a considerable amount of material is saved, not counting the labour saved in foundations, connections etc. As foundations have to grow roughly with the square of the load (e.g. 2 x load = 2 x area x 2 x depth) the economy is not always achieved. The essence lies in reducing span because, broadly, columns are cheaper than beams. This agrees also with the economical concept of heavily loaded beams.

Continuity in structure

In the preceding chapters, we have seen how bending moments grow with span for fixed loads, and with the square of the span for

110

continuous loads. We also noted how the material needed increases accordingly. We have seen, too, how deflection can grow at even more alarming rates. So both bending moments and deflection must be reduced in the interest of economic design.

It was shown in Chapter 7 how pre-stressing, by introducing initial negative moments to counteract the loading effect, was one way of reducing the effect of bending moments. Initial bending moments can also be created by fixing moments (Fig. 190).

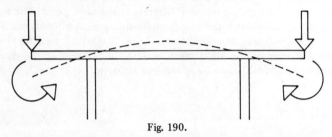

Fig. 190.

These are similar to the moments produced by cantilevers. Figure 191 shows a structural way of achieving counteracting moments. In Fig. 192 is shown the result of overlaying the two moments. Where the moment diagram turns from negative to positive, a point of contra-flexure and therefore, zero moment, occurs.

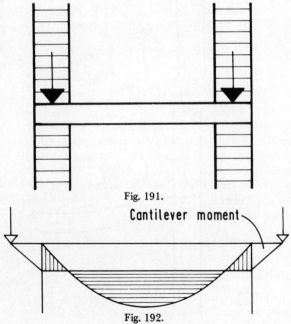

Fig. 191.

Cantilever moment

Fig. 192.

111

A beam can only be regarded as *really* fixed if it remains level at the supports, or if its *slope of deflection* is zero. Figure 193 shows an inadequate, and Fig. 194 an adequate, fixing, producing a negative bending moment called a *fixing moment*.

The total turning effect in one direction must be counteracted by the total turning effect in the other. So the sum of the turning effects must be zero. The turning effect at each point of a beam is represented by the bending moment at this point, which is the ordinate of the bending moment diagram at the same point. The sum of all these ordinates is the area of the bending moment diagram. So, if a "positive" area is counteracted by a "negative" area for zero slope the sum of the areas must be zero, or *the area of the fixing moment* must equal *the area of the free moment*.

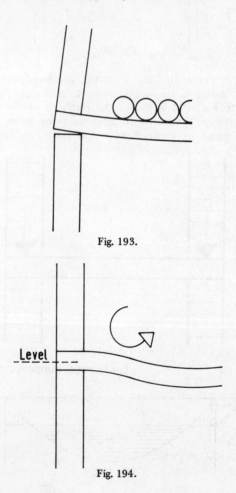

Fig. 193.

Level

Fig. 194.

Figure 195a, b indicates that under these conditions, the free moment is halved. Deflection in this case would be

$$\frac{1}{192}\frac{L^3}{EI}$$

or *one quarter* of that of a freely supported beam

$$\left(\frac{1}{48}\frac{L^3}{EI}\right)$$

For a continuous load, the fixing moment is $WL/12$ and the span moment $WL/24$. These values are derived from the geometry of the parabola the area of which equals that of a rectangle of 2/3 its height. Deflection would then be

$$\frac{1}{384}\frac{WL^3}{EI}$$

or one fifth.

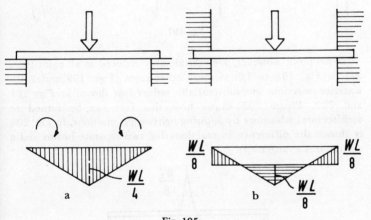

Fig. 195.

Fixity is a normal condition in most framed structures and the accepted corollary of reinforced concrete construction. The degree of fixity depends largely on the rigidity of the supports. This is illustrated in Fig. 196a and b. The conditions here are similar to those in the portal frame (see Chapter 3 and Fig. 197).

Beams *continuous* over several supports produce fixing moments and remain level (or nearly so) over the supports. This produces greatly reduced span moments. Moments over the supports are reduced to a lesser extent. With beams over *three* supports the support moments remain $WL/8$ and the system is of little avail if beams or slabs are kept the same thickness throughout.

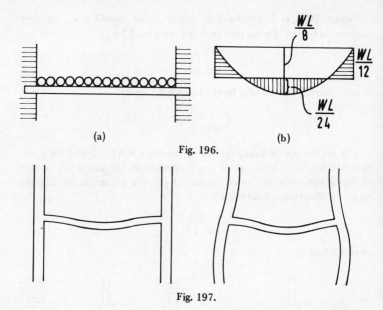

Fig. 196.

Fig. 197.

Deflections, however, are considerably reduced in all cases from 2/5 in Fig. 198 to 1/5 in most other cases (Figs. 199 and 200). Exterior reactions are substantially reduced as shown in Figs 201 and 202. Figure 203 shows how this fact can be turned to architectural advantage by allowing lightweight mullions. In Fig. 204 is shown the difference in reactions for two separate beams and a continuous arrangement.

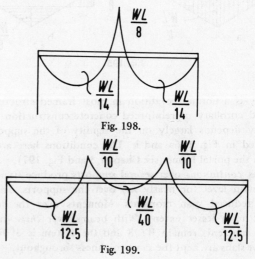

Fig. 198.

Fig. 199.

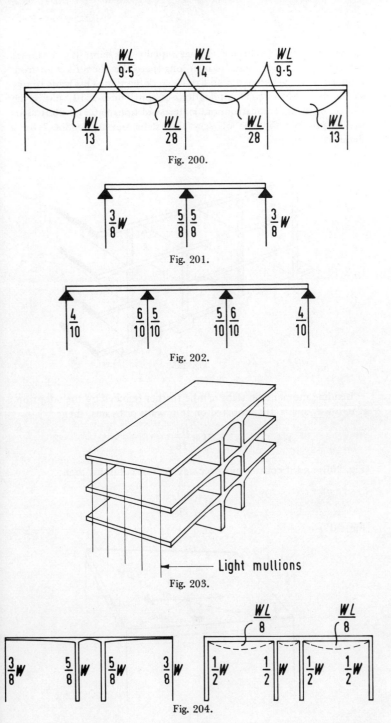

Fig. 200.

Fig. 201.

Fig. 202.

Light mullions

Fig. 203.

Fig. 204.

115

What applies to beams applies equally to slabs. As these are fundamentally uneconomical over long spans they benefit most from reduced bending moments and reduced deflection. This leads to one of the most economical applications of continuous slabs, that is, the use of cross wall construction in terraced housing (Fig. 205). This makes possible the use of very thin slabs (see Appendix 7 for a detailed example).

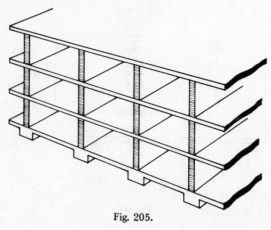

Fig. 205.

Bending moments in slabs can be further reduced by the adoption of two-way spans. If supported on four walls or beams, then:

$$M = \frac{WL}{16}$$

(Fig. 206) and if continuous over a system of columns, then

$$M = \frac{WL}{20}$$

(Fig. 207).

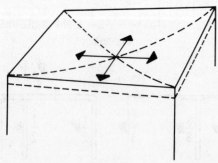

Fig. 206.

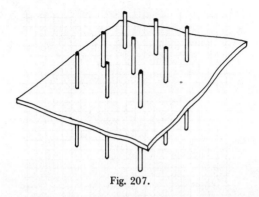

Fig. 207.

The two-way effect is progressively lost as the plan shape deviates from the square. The restraining effect of the edges largely disappears when the long side approaches a length three times that of the short side. With "mushroom head" columns which give additional restraint, the bending moments may be reduced to $WL/40$ (Fig. 208) (CP 114).

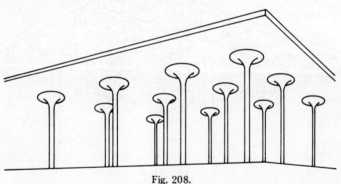

Fig. 208.

Reinforcement has to run in two directions of course and must be increased over the supports to meet local shear (Fig. 209). Where spans become very long, slabs can be replaced by two-way T beams. As there would be a considerable shear load over the columns and relatively little area to meet it, panels adjacent to the columns are filled in solid (Fig. 210). In two-way slabs, stresses are greatest in a *diagonal* direction. Therefore, where the main structural members in a two-way arrangement are few (as in the case of long spans) considerable stress reductions are obtained if the beams follow a diagonal pattern (Figs 211 and 212). This arrangement is called a *diagrid*. For effective two-way action, each member of a diagrid must be supported. Alternatively, the whole system must be supported by a joint edge beam.

117

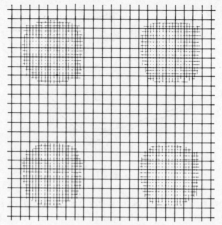

Fig. 209.

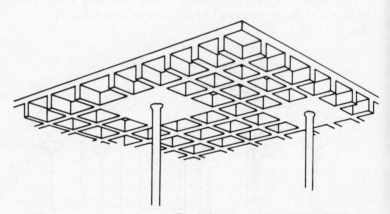

Fig. 210.

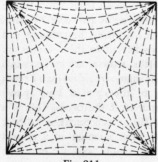

Fig. 211.

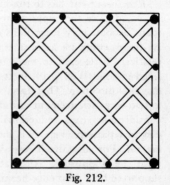

Fig. 212.

Space frames

Where spans are long and loads are light, moments of inertia must be kept high and self-weight low. Thus, lightweight steel trusses can be combined into trussed folded slabs (see p. 67 and Fig. 213). When spanning in one direction only, however, their efficiency is not substantially increased over that of a series of closely spaced trusses.

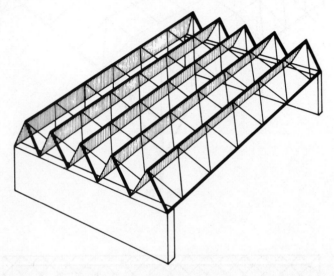

Fig. 213.

But if ridges and bottom chords are linked, the system becomes two-way and if used over spans that are near square, their action resembles that of the two-way slab. The arrangements then becomes suitable for spans of 20 m and upward (Fig. 214).

As for slabs, the stress resultants run prevailingly on the diagonal and the best arrangement therefore is on the diagrid principle (Fig. 215). Bending moments are further reduced by cantileverage (see p. 35). As in the case of the mushroom columns, spreader heads to the supports give additional fixity and reduce local shear loads (Fig. 208).

Essentially, the space frame is *not* an economical concept, because it represents a series of closely spaced and lightly loaded beams carrying over long spans. While its overall depth is less than that required by an economical arrangement of trusses, it is still substantial. Furthermore, a great deal of material is required, which makes the space frame expensive and difficult to maintain. But over really long spans, if a flat roof is essential with a minimum of supports, it can be an economical form of structure.

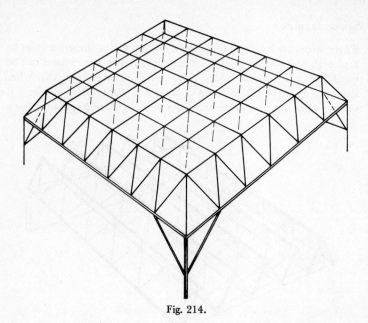

Fig. 214.

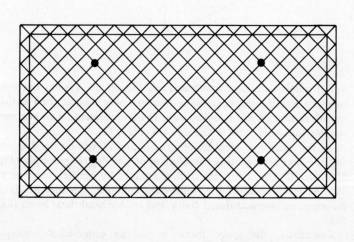

the space to new additional fluity and to .... the load there being high.

... condition, the shear strain is not an essential amount ...

that superimposed ... even .... as ... an inimum moment, it is ...
obtained. Even ... should pass ... of in ... is required, which
... the ... could be ... superstructure ... for the ... is ... is ... But even
being ... there is a ... to present it will be a ... in amount of
... or ... could be ... . not ... is ... in ... structure.

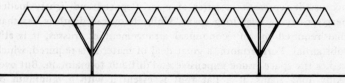

Fig. 215.

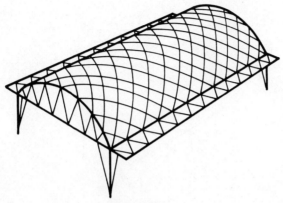

Fig. 216.

Bending moments may be further reduced by introducing the third dimension into the space slab by pitching, arching etc. (Fig. 216), but a great deal of standardization is lost by such arrangements.

Further information on space frames will be found in another volume of this series, "Roofs" by Roy E. Owen, ARIBA.

121

9 Plastic theory and the load factor method of calculation

The method of analysis explained in Chapter 4 was based entirely on the *elastic behaviour* of materials and the concept of *maximum stress*. It was shown that materials do not necessarily collapse when the maximum stress is reached (see the graph in Fig. 90). In the case of steel, as shown, a plastic range comes into play beyond 230 N/mm^2 where deformation continues without appreciable increase in stress.

In a beam stressed in this way hinges would develop. A beam on *two* supports would then collapse as shown in Fig. 217. A two pin frame, on the other hand, would become virtually a *three* pin frame and would not collapse until a further hinge developed (Figs 218 and 219). The safe load, therefore, would be a fraction not of a load

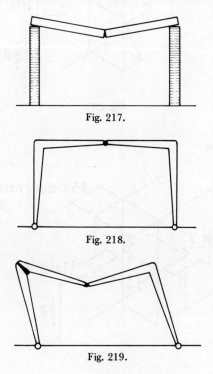

Fig. 217.

Fig. 218.

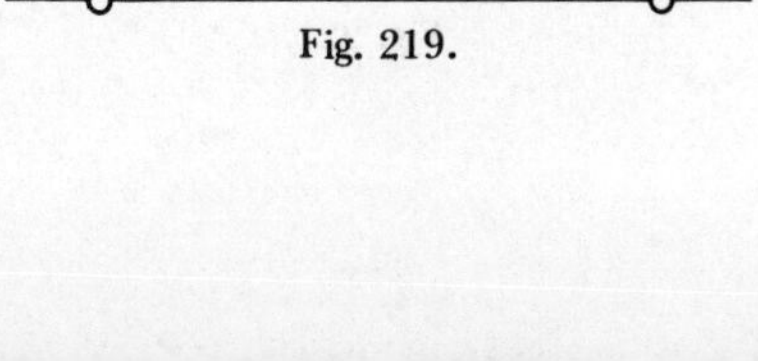

Fig. 219.

which would produce a maximum stress in the outer fibres of a member, but of a load which would actually produce collapse. The relationship between

$$\frac{\text{Collapse Load}}{\text{Working load}}$$

is called the *load factor*.

A hinge cannot develop until the limit of elasticity is reached over the *entire section* (not only the extreme fibres). This leads to a different stress picture. Figure 220 shows the elastic stress distribution and its lever arm, while Fig. 221 shows the stress distribution of the plastic stage with its lever arm.

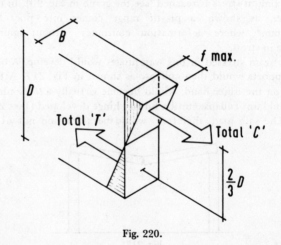

Fig. 220.

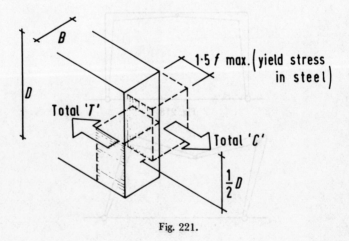

Fig. 221.

As always the resisting moment is produced by the total compression or total tension multiplied by their lever arm from each other. Thus, in Fig. 221

$$M_r = B \times \frac{D}{2} \times \frac{D}{2} \times 1.5 f_{max} \text{ (elastic)}$$
$$= \frac{BD^2}{4} \times 1.5 f_{max} \text{ (elastic)}$$

The expression $BD^2/4$ is called the *plastic modulus* and is 1½ times the elastic modulus $(BD^2/6)$.

So: $M = 1.5 \times 1.5 = 2.25 \times f_{max}$

or, since bending moments are directly proportional to the load, collapse would occur with a load 2.25 times the elastic load.

This figure is therefore an indication of the load factor. The *actual* load factor depends on a number of considerations, of which the most important and obvious is the working load actually allowed. This is, of course, much higher than the load leading to the development of maximum fibre stress.

Further, the load factor varies with the shape of the section involved and the structural situation. For instance, the load that would lead to collapse bears a different relation to the working load in the case of a beam on two supports or on one forming part of a portal frame (see Figs 218 and 219).

The influence of the *shape* of the section is shown in Fig. 222 and Fig. 223. The placing of the bulk of the material near the outer edge is now of less importance than it would be under elastic conditions, since the *total* section plays a larger part, and due to this and to the higher stress, contributes a larger stress content. This is partially counteracted by the fact that the lever arm is shorter.

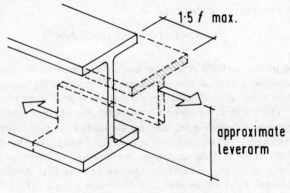

Fig. 222.

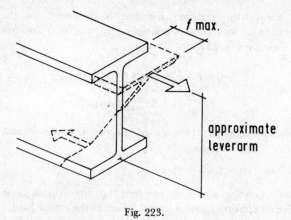

f max.

approximate
leverarm

Fig. 223.

This introduces the concept of the *shape factor*. In the case of the rectangle the plastic Z was 1½ times as large as the elastic Z.

$$\frac{BD^2}{4} = 1\frac{1}{2} \times \frac{BD^2}{6}$$

This figure, "1½", is called the *shape factor*.

In the case of *I* or similar sections, the proportion between plastic and elastic Z is only about 1.15. Therefore, the load factor for a yield stress of 1.5 times the working stress would only be 1.15 x 1.5 or about 1.75. The load factor also varies between different materials. For example, the plastic behaviour of concrete is shown in Fig. 224. The stress at which a test cube would collapse is the "cube strength" which, in the case of good quality concrete, is of the order of 31.5 N/mm² (4500 lb/in.²). However, plasticity would begin at a value of about two thirds of this, namely 21 N/mm² (3000 lb/in.²). The relevant Code of Practice recommends a load factor of 1.8, so that the resulting working stress is

$$\frac{1}{1.8} \times 21 = 11.7 \text{ N/mm}^2 \text{ (1670 lb/m}^2\text{)}$$

The plastic stress distribution is again assumed to be a rectangle extending to a neutral axis, the depth of which varies according to the percentage of steel used, thus resulting in lever arms of varying depth. The stress distribution also depends on whether steel is used in compression (which is required by the Code in certain cases of low cube strength) or on shallow depths of beams. In the case of balanced sections (that is, where the amount of steel is able to produce exactly the same resistance moment as that produced by the concrete in compression), the lever arm equals 0.75 x D. This leads to resisting moments substantially higher than those based on

126

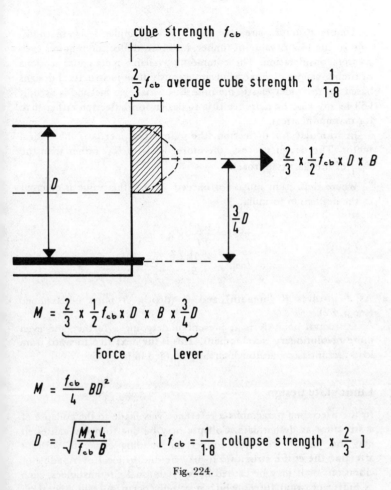

cube strength f_{cb}

$\frac{2}{3} f_{cb}$ average cube strength $\times \frac{1}{1 \cdot 8}$

$\frac{2}{3} \times \frac{1}{2} f_{cb} \times D \times B$

$M = \frac{2}{3} \times \frac{1}{2} f_{cb} \times D \times B \times \frac{3}{4}D$

$\underbrace{\phantom{\frac{2}{3} \times \frac{1}{2} f_{cb} \times D \times B}}_{\text{Force}} \quad \underbrace{\phantom{\times \frac{3}{4}D}}_{\text{Lever}}$

$$M = \frac{f_{cb}}{4} BD^2$$

$$D = \sqrt{\frac{M \times 4}{f_{cb}\, B}} \qquad [\; f_{cb} = \frac{1}{1 \cdot 8} \text{ collapse strength} \times \frac{2}{3} \;]$$

Fig. 224.

the elastic theory; shallower sections are involved and the question of deflection thus becomes the main design criterion (see P. 000). Very often, the resulting depths are below the minima required in the Code of Practice (CP 114) (see Figs 111 and 112) and only the steel content has to be checked. This will be higher, due to a lever arm reduced twice, namely less depth as such and only 0.75 of this lesser depth (see p. 67 and Appendix 7 for a computation of the steel area). The Code referred to gives tables of lever arms for T and L beams, but usually, these are assumed as 0.75 x D (measured to the centre of the slab).

The interesting fact emerges that in the plastic method of concrete design, the concept of relative E values between steel and concrete no longer matters, the concrete being used far more realistically in accordance with its actual behaviour under load.

127

Timber structures are now based on very similar design methods. Due to the low E value of timber, however, deflection requires even greater consideration. The commonly available rectangular sections of timber can be used more advantageously than in structural designs based on the older elastic method. But whichever method is used, it will in any case be more realistic to design for deflection rather than for maximum stress.

In formulae for deflection, the moment of inertia is the critical factor. The design process, therefore, leads to I_{req} rather than the Z_{req} of the elastic approach.

Where deflection must not exceed $\frac{1}{325} L*$ this value is equated to the deflection formula:

$$\frac{1}{325} L = \frac{WL^3}{EI} \times c$$
$$I_{req.} = \frac{325 WL^2}{E} \times c*$$

As I involves B (breadth) and D (depth, B must be assumed (see p. 62).

Structural analysis is at present undergoing a further and even more revolutionary development. This is the next step forward from load-factor design, although already implied in it.

Limit state design

In the preceding paragraphs a reference was made to the collapse of a structure as design data. Collapse may be due to many causes: it may be partial, confined to certain building elements without affecting the entire structure; it may be due to local overloading or vibration; or it may be caused by very unusual circumstances, such as hurricane wind forces which may only occur once in a century. Engineers are already using their discretion when they assess certain loading conditions and to give up the concept of the maximum permissible fibre-stress as the only design criterion, wherever it might occur, was the first step in this direction.

The next step was to design beam and column sections on the plastic basis, utilizing the entire section rather than only the extreme fibres.

The latest approach to design attempts to meet all the conditions which can lead to the *limit state*, either in the form of collapse or extreme strain (which in practice would lead to excessive deflection), by codifying the loads and stresses which would lead to these limiting states.

* In the case of steel and timber see also footnote on p. 51.

Instead of laying down hard and fast values, the latest limit state codes are based on statistics, for example, the frequencies of certain windloads or variations in tests. These are called characteristic loads, and are augmented by factors allowing for possible deviations from such values. This leads to even higher local stress allowances called characteristic stresses, in certain situations reduced by certain safety factors and higher deflections for entire building types. So *limit state design* takes into account the structural behaviour of entire buildings and the interrelation of their elements.

Generally speaking the limit state approach represents a far more realistic attitude to structural design than the elastic theory, in which not only substantial parts of sections, but even whole parts of the building, remain understressed.

In spite of its great advantages, however, the limit state approach is often not applicable because excessive deflection frequently occurs before the permitted stresses are reached. This is almost invariably the case with beams on two supports.

Almost without exception, therefore, the method finds its best application in statically indetermined situations where deflection is substantially less and spare hinges can develop. This is usually the case in situations of continuity, such as fixed end beams and portal frames with fewer than three hinges. In timber it is structures such as these which open up the most promising possibilities. The method has already been codified for reinforced concrete, where, due to the homogeneity of concrete structures, it finds its widest application.

The load factor method and limit state design does *not* affect the overall geometry of layouts as span-load relations remain fundamentally the same and the shapes of sections are only marginally affected. However, it leads generally to more *slender* sections, and portal frames can have the same section throughout

Although I sections have less favourable shape factors (Figs 222 and 223) and hence load factors less than those of rectangular sections (about 1.75), the total material used will still be less in most cases, resulting in more elegant structures.

The plastic theory does not invalidate the importance of the elastic theory which is derived from observable structural behaviour within the elastic range, that is, long before breakdown occurs. Elasticity is a very noticeable behaviour in all materials. It permits the calculation of deflection and of pressures due to eccentric loading. Also, the analysis of pre-stressing is derived from this. The parabolic distribution of shear stresses over homogeneous cross sections is the result of elastic stress distribution, and the concept of all skin structures is the outcome of elastic stress considerations, whether they be the large units of industrialized building components or the stressed skin of plywood floor panels.

10 The essence of structure

Equilibrium has been shown to be the basis of all structural concepts. Whatever we build, static equilibrium has to be established to meet the forces of Nature which try to disrupt our structures. The need to enclose space prevents us from always guiding forces along the lines of direct equilibrium and all structural problems derive from this. Funicular lines (catenaries) come closest to the ideal and so does any structural form whose shape approaches the catenary. The most elementary form of structure is post and beam. This deviates most from our basic concept, and loads on such a simple structure deviate more from an equilibrating reaction if they are concentrated, than if they are distributed. From this elementary truth derives one of the fundamentals of the geometry of layout; namely that wherever possible, point loads should be avoided.

Bending moments which are the result of the detour of forces determine the shape of sections and structural elements and the *moment of inertia* is an indication of the distribution of the material. The further that material is spread away from a centre of gravity, the greater the lever arm available to meet external moments and resisting moments in sections grow with the square of the spread. In bridging members, the lever arms are created by depth and in supporting members by spread all round, for example, in tubes.

High tensile materials such as steel have made it possible to increase spans because bending moments require couples of compression and tension to produce the equilibrating moments.

Great progress has been made in the construction of pure tensile structures such as suspension bridges and suspended roofs, in which all stresses are concentrated within thin skins only. The compensating compressive stresses are taken up by the ground, so that the entire depth of a building can become the "lever arm".

Not much further development can be expected in this direction so long as E values remain relatively low and structures deflect and deform long before their maximum stresses are reached and, therefore, before full use can be made of their strength. Skin structures in plastics (such as glass fibre reinforced resins) are used for boat hulls and building components of moderate dimensions. No great changes in building structures can be anticipated until materials become available which combine much higher strength than at

131

Fig. 225. Stonehenge

present, with higher E values and lower weight. Certain carbon fibres have this quality but at the moment their cost of production places them out of range for building. On the other hand steel was too costly for builders until new processes of production led to a structural revolution and the price of steel dropped to one-ninth during the reign of one Queen!

The fundamental change that has occurred in building — and hence in the understanding of structural behaviour — is the approach to building in the form of continuous structures. A great deal of construction is no longer the laborious piling up of small components one on top of another. Even here, modern toy sets produce continuity by pegging or adhesives. Like plants and animals, ourselves included, buildings have become organic entities. Nature does not know a loose piling up, but recognizes only organic growth. Examples of this are three dimensional frames, with protective skins or continuous shells which, as with crustaceans or insects, afford protection and at the same time provide structural strength. This is all part of the great human endeavour to come to terms with Nature and her laws and in doing so we come closer to the forms of Nature herself. The load factor method leading to limit state design has given us one more tool for the rational assessment of the combined action of the elements of which buildings are composed.

A long way from Stonehenge!

Reading list

H. Werner Rosenthal, *Structural Decisions*, Chapman and Hall.

W. Fisher Cassie & Napper, *Structure in Building*, Architectural Press.

Zuk, *Concepts of Structure*, Reinhold.

Salvadori Heller, *Structure in Architecture*, Prentice Hall Inc.

Torroja, *Philosophy of Structure*, University of California Press.

H. Seymour Howard Jr., *Structure, an Architect's Approach*, McGraw-Hill.

Curt Siegel, *Structure and Form*, Crosbie Lockwood.

W. Morgan, *Students' Structural Handbook*, Butterworth.

J. E. Gordon, *The New Science of Strong Materials*, Penguin.

Frei Otto, *Tensile Structures* Volumes I and II, M.I.T. Press, Cambridge, Mass.

Z. Makowski, *Steel Space Structures*, Michael Joseph.

British Standard Codes of Practice:
CP 112 *The Structural Use of Timber.*
CP 114 *The Structural Use of Reinforced Concrete in Buildings.*
CP 115 *Structural Use of Prestressed Concrete in Buildings.*
CP 116 *Structural Use of Precast Concrete.*

British Standard 449 *Use of Structural Steel in Building.*

Timber Research and Development Association (TRADA), *Timber Frame Housing Design Guide.*

Appendix 1
SI conversion table

Imperial	SI	Metric	
1 in.		25.4 mm	
1 ft		304.8 mm	Length
3.28 ft		1 m	
1 in.^{2}		645.2 mm^2	
1 ft^2		0.093 m^2	Area
10.7 ft^2		1 m^2	
1 lb		0.454 kg	
2.2 lb		1 kg	Mass
1 ton		1016.05 kg	(weight)
0.982 tons		1 t	
1 ton/ft^2	107.2 kN/m^2		
1 ton/in^2	15 480 kN/m^2		Live loads
	15.48 N/mm^2		
20.88 lb/ft^2	1.0 kN/m^2		
1 lb f	4.448 N		
0.225 lb f	1 N		Force
1 ton f	9964 N		moments
	9.9 kN		
1 ton x in.	0.253 kN x m		
1 lb/in.^{2}	0.0069 N/mm^2		
1 ton/in.^{2}	15.45 N/mm^2		
144.93 lb/in.^{2}	1 N/mm^2		Stress
1000 lb/in.^{2}	6.9 N/mm^2		
0.065 ton/in.^{2}	1 N/mm^2		

Appendix 2
Calculation of a roof truss

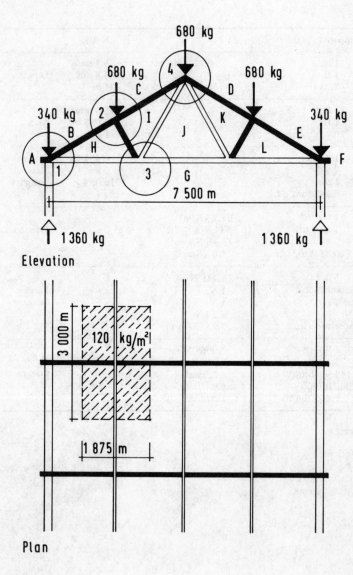

680 kg

680 kg   4   680 kg
C   D

340 kg   340 kg
2   I   K
B   E
H   J
A   3   G   F
1   L
7 500 m

1 360 kg   1 360 kg

Elevation

3 000 m
120 kg/m²

1 875 m

Plan

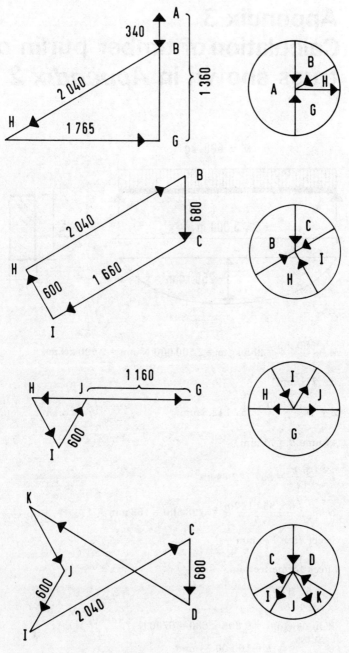

Forces scale : 1mm = 40 kg

137

Appendix 3
Calculation of timber purlin on truss shown in *Appendix 2*

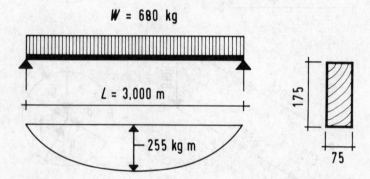

$$M = \frac{680 \times 3}{8} = 255 \text{ kg m} = 2\,500\,000 \text{ N mm} = 2\,500 \text{ kN m}$$

$$f = 7 \text{ N/mm}^2$$

$$Z = \frac{2\,500\,000}{7} = 357\,142.8 \text{ mm}^3$$

Assume $B = 75$ mm

$$Z = \frac{75D^2}{6} = 357\,142.8 \text{ mm}^3$$

$$D = \sqrt{\frac{6 \times 357\,142.8}{75}} = \sqrt{28\,500} = 168 \text{ mm}$$

Select 75 x 175 mm

Check for deflection

$$d = \frac{5}{384} \frac{WL^3}{EI}$$

W in newtons $= 9.806 \times 680 = 6700$ N

$$E = 10\,500 \text{ N/mm}^2$$

$$I = 34\,200\,000 \text{ mm}^4$$

138

$$d = \frac{5 \times 6700 \times 3000^3}{384 \times 10\,500 \times 34\,200\,000}$$

$$= \frac{5 \times 6.7 \times 3^3}{3.84 \times 1.05 \times 3.42} \frac{10^3 \times 10^9}{10^2 \times 10^4 \times 10^7}$$

$$= \frac{65.6}{10} = 6.56\,\text{mm}$$

Permissible deflection $\frac{1}{325}$ of span

$$\frac{6.56}{3000} = \frac{1}{460}\,\text{span}$$

$$I = \frac{75 \times 175^3}{12} = 34.2 \times 10^6\,\text{mm}^4$$

Appendix 4
Calculation of a timber box girder

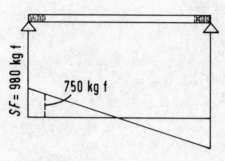

Shear diagram

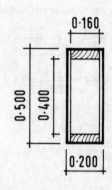

Cross-section

Loading 120 kg/m^2 $W = 1800$ kg

$$M_{req.} = \frac{W \times L}{8} = \frac{1800 \times 10}{8}$$

$$= 2250 \text{ kg m}$$

or 22.0 kN m

Check M possible for $f = 6.3$ N/mm^2

$$I = \frac{1}{12}(BD^3 - bd^3) = \frac{1}{12}(200 \times 500^3 - 160 \times 400^3)$$

$$= \frac{1}{12}(2 \times 5^3 - 1.6 \times 4^3) \times 10^8 = 12.3 \times 10^8$$

$$Z = \frac{I}{Y} = \frac{12.3 \times 10^8}{250} = 0.4925 \times 10^8 = 4\,925\,000 \text{ mm}^3$$

$$M_{poss.} = 4\,925\,000 \times 6.3 = 31\,000\,000 \text{ mN} \times \text{m}$$

$$= 3150 \text{ kg} \times \text{m}$$

adequate

Check for shear

$$S_{max} = \frac{1800}{2} \, kgf$$

$$\text{Shear stress} = \frac{S}{A_{flanges}}$$

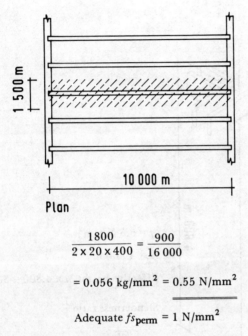

Plan

$$\frac{1800}{2 \times 20 \times 400} = \frac{900}{16\,000}$$

$$= 0.056 \, kg/mm^2 = 0.55 \, N/mm^2$$

$$\text{Adequate } fs_{perm} = 1 \, N/mm^2$$

Ends to be strengthened against buckling.

Appendix 5
Calculation of a slender timber column

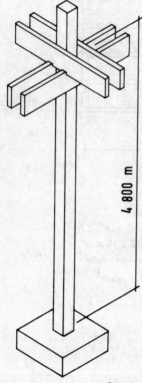

$f = 5.5 \text{ N/mm}^2$

$H = 4.800 \text{ m}$

Effective height $\frac{3}{4} \times 4.800 = 3.600 \text{ m}$

Slenderness ratio $\frac{L}{r}$

$r = \sqrt{\dfrac{I}{A}} = \sqrt{8.33} = 28.8 \quad \dfrac{L}{r} = 125$

f_{perm} : (Table Fig. 151) interpolated
between 120 and
130 factor = 0.38

$f_{\text{perm}} = 0.38 \times 5.5 \text{ N/mm}^2 = 2.09 \text{ N/mm}^2$

$P_{\text{permissible}} = 2.09 \times 100^2 = 20\,900 \text{ N} = 2130 \text{ kg}f$

$I = \dfrac{100^4}{12} = 8\,333\,000 \text{ mm}^4$

From Euler:

$$P = \frac{\pi}{L^2} EI \qquad E \text{ (assumed for grade}$$
$$S_2 = 10\ 300\ \text{N/mm}^2)$$

$$P = \frac{\pi}{3600^2} \times 10\ 300 \times 8\ 333\ 000\ \text{N}$$

$$= \frac{\pi}{3.6^2} \times 1.03 \times 8.3\ \frac{10^4 \times 10^6}{10^6} = 20\ 800\ \text{N}$$

$$= 2120\ \text{kg}f$$

Appendix 6
"Kern" dimensions for hollow shapes

1. Squares

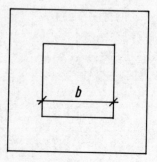

$$e = \frac{Z}{A}$$

$$Z = \frac{I_B - I_b}{B/2} = \frac{1}{12} \frac{B^4 - b^4}{B} \times 2$$

$$A = B^2 - b^2$$

$$\frac{Z}{A} = \frac{1}{6B} \frac{B^4 - b^4}{B^2 - b^2}$$

$$= \frac{1}{6B} (B^2 + b^2)$$

For very thin walls

b approaches B

$$e = \frac{1}{6B} (B^2 + B^2) = \frac{B}{3}$$

which is 2 x eccentricity of solid

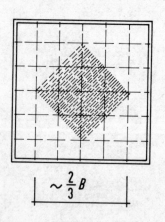

$\sim \frac{2}{3} B$

2. Tubes

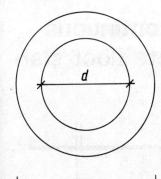

$$Z = \frac{I_D - I_d}{D/2} = \frac{\pi}{64} \frac{D^4 - d^4}{D/2}$$

$$A = \frac{\pi(D^2 - d^2)}{4}$$

$$\frac{Z}{A} = \frac{1 \times 2}{16D} \times \frac{D^4 - d^4}{D^2 - d^2} = \frac{1}{8D} \times (D^2 + d^2)$$

For very thin walls

d approaches D

$$e = \frac{1}{8D} \times 2D^2 = \frac{D}{4}$$

which is 2 × eccentricity of the solid.

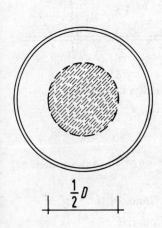

$\frac{1}{2}D$

Appendix 7
Calculation of a continuous reinforced concrete floor slab

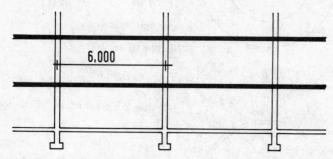

Live load 5 kN/m^2 [100 lb/ft^2]

Dead load $\underline{3.6 \text{ kN/m}^2}$ [72 lb/ft^2] assuming a slab 150 mm thick

  8.6 kN/m^2

$W = 6 \times 8.6 = 51.6$ kN

$M = \dfrac{51.6 \times 6}{10} = 31.0$ kN m $M = \dfrac{P_{CB}}{4} BD^2$ [load factor method]

$D = \sqrt{\dfrac{31.0}{1750 \times 1}} = \sqrt{0.0177}$ $P_{CB} = 7$ N/mm^2 = 7000 kN/m^2

  $= \underline{0.130 \text{ m}}$ $M = 1750 \, BD^2$

Steel

$M = A_s \times f_s \times$ lever

Lever arm $= \dfrac{3}{4}D = \dfrac{3}{4} \times 0.130 = 0.097$ m

  $f_s = 210\,000$ kN/m^2

$31.0 = A_s \times 210\,000 \times 0.097$

$A_s = \dfrac{31.0}{21\,000} = 0.00\,147$ m^2 = 2290 mm^2

Assume 10 bars 147 mm^2 each $\dfrac{\pi D^2}{4} = 147$

Use 10 bars 14 mmϕ or nearest $D = 14$ mm.

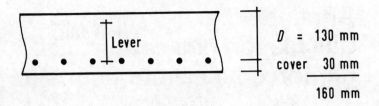

D = 130 mm

cover   30 mm

160 mm

Index

149

151

154